世界园林
WORLDSCAPE

No.1 2014

中国林业出版社

本辑主题：2013年（第四届）"园冶杯"风景园林国际竞赛专辑

THEME: SPECIAL ISSUE OF THE 2013 FOURTH ANNUAL "YUAN YE AWARD" INTERNATIONAL LANDSCAPE ARCHITECTURE COMPETITION

2014年（第五届）"园冶杯"

The 2014 Fifth Annual "Yuan Ye Award" Inter

2014年（第五届）"园冶杯"风景园林(毕业作品、论文)国际竞赛是由中国建设教育协会和中国花卉园艺与园林绿化行业协会主办，中国风景园林网和《世界园林》杂志社承办，在风景园林院校毕业生中开展的一项评选活动。

时间安排
报名截止日期：2014年4月30日

资料提交截止日期：2014年6月10日

参赛资格
应届毕业生（本科、硕士、博士）

参赛范围
风景园林及相关专业的毕业作品、论文均可报名参赛

竞赛分组
竞赛设置四类：风景园林设计作品类、风景园林规划作品类、园林规划设计论文类、园林植物研究论文类。每类设置两组：本科组和硕博组

Time schedule
The application deadline is April 30,2014;Submission deadline is June 10, 2014

Qualification
This year's graduates (Bachelor,Master,Doctor)

Competition content:
Landscape architecture and the related specialized graduation work, or paper

Types
There are four types:The design works of landscape architecture,the planning works of landscape architecture, the design and planning papers of landscape architecture,and the papers of garden plants. Every type is set into two groups: the undergraduate group,the master and doctor group.

风景园林（毕业作品、论文）国际竞赛
nal Landscape Architecture Graduation Project/Thesis Competition

The 2014 Fifth Annual "Yuan Ye Award" International Landscape Architecture Graduation Project is hosted by China Construction Education Association, Chinese Flowers Gardening and Landscaping Industry Association, and is undertaken by China Landscape Architecture network and *Worldscape*. It is a competition among graduates in landscape architecture schools.

地址：北京市海淀区三里河路17号甘家口大厦1510
邮编：100037 电话：(86) 010-88364851
传真：(86) 010-88365357
学生咨询邮箱：yyb@chla.com.cn
院校咨询邮箱：messagefj@126.com
官方网站：http://www.chla.com.cn

Address: 1510#, Ganjiakou Building, No.17, Sanli River Road, Haidian District, Beijing.
Post Code: 100037 **Tel:** 010-88364851
Fax: 010-88365357
Student Advisory E-mail: yyb@chla.com.cn
College Advisory E-mail: messagefj@126.com
Website: http://www.chla.com.cn

东方园林股票代码：002310

城市景观生态系统运营商

1992-2012

东方园林
20年
2000人
20座

城市景观艺术品
OrientLandscape
Urban landscape art

OL
Orient Landscape

中国园林第一股
全球景观行业市值最大的公司
中国A股市场建筑板块、房地产板块前十强
城市景观生态系统运营商

北京奥林匹克公园中心区景观
北京通州运河文化广场
首都机场T3航站楼景观
北京中央电视台新址景观
苏州金鸡湖国宾馆、凯宾斯基酒店景观
苏州金鸡湖高尔夫球场
上海佘山高尔夫球场
上海世博公园
海南神州半岛绿地公园
山西大同新城中央公园文瀛湖
湖南株洲新城中央公园神农城
辽宁鞍山新城景观万水河
辽宁本溪新城中央公园
河北衡水衡水湖及滏阳河景观
山东滨州生态景观系统及新城中心景观
浙江海宁生态景观系统
山东淄博淄河景观系统
河北张北风电基地及两河景观带
山东济宁微山湖及任城区中央景观
山东烟台夹河景观系统及特色公园

世界园林
WORLDSCAPE

主办单位	亚洲园林协会	
联合主办	国际绿色建筑与住宅景观协会	

（按姓氏字母顺序排名）

总　编　王小璘（台湾）
副总编　包满珠　陈弘志（香港）　李敏　沈守云　王浩　朱育帆
顾问编委　Peter Latz（德国）　凌德麟（台湾）　罗哲文　Peter Walker（美国）
编委会
常务编委　Jack Ahern（美国）　Henri Bava（法国）　曹南燕　陈溱溱　高翅
Christophe Girot（瑞士）　Karen Hanna（美国）　何友锋（台湾）　贾建中
况　平　Eckart Lange（英国）　李如生　李　雄　李炜民　Patrick Miller（美国）
欧圣荣（台湾）　强　健　Phillippe Schmidt（德国）　Alan Tate（加拿大）
王庚飞　王良桂　王向荣　谢顺佳（香港）　杨重信　喻肇青（台湾）
章俊华　张　浪　赵泰东（韩国）　朱建宁　朱育帆
编　委　白祖华　陈其兵　成玉宁　杜春兰　方智芳（台湾）　黄哲　简仔贞（台湾）
金晓玲　李春风（马来西亚）　李建伟　刘纯南　刘庭风　罗清吉（台湾）
马晓燕　蒙小英　Hans Polman（荷兰）　邱坚珍　瞿　志　宋红红　王鹏伟
王秀娟（台湾）　吴静宜（台湾）　吴雪飞　吴怡彦（台湾）　夏海山　夏　岩
张莉欣（台湾）　张青萍　周武忠　周应钦　朱　玲　朱卫荣　张新宇　赵晓平
编　辑　傅　凡　高　杰　胡雅静　焦子源　马一鸣　覃　慧（台湾）　宋焕芝　张　安
张红卫　赵彩君　郑晓笛
外文编辑　何友锋（台湾）　Charles Sands（加拿大）（主任）　Trudy Maria Tertilt（德国）
谢顺佳（香港）　朱　玲
版式设计　王　薇
责任编辑　田　娟
广告发行　010-88364851

地　址
北　京　北京市海淀区三里河路17号甘家口大厦1510
邮编：100037　电话：86-10-88364851
传真：86-10-88365357　邮箱：worldscape@chla.com.cn
香　港　香港湾仔骆克道315-321号骆中心23楼C室
电话：00852-65557188　传真：00852-31779906
台　湾　台北书局
台北市万华区长沙街二段11号4楼之6
邮编：108　电话：886+2-23121566，
传真：886+2-23120820　邮箱：nkai103@yahoo.com.tw
封面作品　成都保利皇冠假日酒店（图片来源：EDSA Orient）

图书在版编目（CIP）数据

世界园林．2013年（第四届）"园冶杯"风景园林国际竞赛专辑：汉英对照/亚洲园林协会主编．
--北京：中国林业出版社，2014.4
ISBN 978-7-5038-7433-8
Ⅰ．①世… Ⅱ．①亚… Ⅲ．①园林－介绍－世界－汉、
英 Ⅳ．①TU986.61
中国版本图书馆CIP数据核字（2014）第064723号

中国林业出版社
责任编辑：李　顺　纪　亮
出版咨询：（010）83223051

出　版：中国林业出版社（100009 北京西城区德内大街刘海胡同7号）
印　刷：北京博海升印刷有限公司
发　行：中国林业出版社
电　话：（010）83224477
版　次：2014年3月第1版
印　次：2014年3月第1次
开　本：889mm×1194mm 1／16
印　张：12.5
字　数：200千字
定　价：80.00RMB（30USD，150HKD）

Host Organizations
　　Asian Landscape Architecture Society
Co-host Organization
　　International Association of Green Architecture and Residential Landscape
Editor-in-Chief
　　Xiaolin Wang（Taiwan）
Deputy Editors
　　Manzhu Bao　Hungchi Chen (HongKong)　Min Li　Shouyun Shen　Hao Wang　Yufan Zhu
Consultants
　　Peter Latz (Germany)　Delin Ling (Taiwan)　Zhewen Luo　Peter Walker (USA)
Editorial Board
Managing Editors
　　Jack Ahern (USA)　Henri Bava (France)　Nanyan Cao　Zhenzhen Chen　Chi Gao
　　Christophe Girot (Switzerland)　Karen Hanna (USA)　Youfeng He (Taiwan)　Jianzhong Jia
　　Ping Kuang　Eckart Lange (England)　Rusheng Li　Xiong Li　Weimin Li　Patrick Miller (USA)
　　Shengrong Ou (Taiwan)　Jian Qiang　Phillippe Schmidt (Germany)　Alan Tate (Canada)
　　Gengfei Wang　Lianggui Wang　Xiangrong Wang　Shunjia Xie (HongKong)
　　Chongxin Yang (Taiwan)　Zhaoqing Yu (Taiwan)　Junhua Zhang　Lang Zhang
　　Taedong Jo (Korea)　Jianning Zhu　Yufan Zhu
Senior Editors
　　Zuhua Bai　Qibing Chen　Yuning Cheng　Chunlan Du　Zhifang Fang (Taiwan)
　　Zhe Huang　Yuzhen Jian (Taiwan)　Xiaoling Jin　Chunfeng Li (Malaysia)　Jianwei Li
　　Chunqing Liu　Tingfeng Liu　Qingji Luo (Taiwan)　Xiaoyan Ma　Xiaoying Meng
　　Hans Polman (Netherlands)　Jianzhen Qiu　Zhi Qu　Yuhong Song　Pengwei Wang
　　Xiujuan Wang (Taiwan)　Jingyi Wu (Taiwan)　Xuefei Wu　Yiyan Wu (Taiwan)　Haishan Xia
　　Yan Xia　Lixin Zhang (Taiwan)　Qingping Zhang　Wuzhong Zhou　Yingqin Zhou　Ling Zhu
　　Weirong Zhu　Xinyu Zhang　Xiaoping Zhao
Editors
　　Fan Fu　Jie Gao　Yajing Hu　Ziyuan Jiao　Yiming Ma　Hui Qin (Taiwan)　Huanzhi Song
　　An Zhang　Hongwei Zhang　Caijun Zhao　Xiaodi Zheng
Foreign Language Editors
　　Youfeng He (Taiwan)　Charles Sands (Canada, Director)　Trudy Maria Tertilt (Germany)
　　Shunjia Xie (Hongkong)　Ling Zhu
Layout Design　Wei Wang
Editor-In-Charge　Juan Tian
Advertising & Issuing　010-88364851

Corresponding Address
　　Beijing　1510 Room, Gan Jia Kou Tower, NO. 17 San Li He Street, Haidian District, Beijing P.R.C
　　Code No. 100037　Tel：86-10-88364851　Fax：86-10-88365357　Email：worldscape@chla.com.cn
HongKong
　　Flat C,23/F,Lucky Plaza,315-321 Lockhart Road, Wanchai, HONGKONG
　　Tel：00852-65557188　Fax：00852-31779906
Taiwan　Taipei Bookstore
　　6#,the 4th Floor , Changsha Street Section No.2, Wanhua District , Taipei Code No. 108
　　Tel：886+2-23121566　Fax：886+2-23120820　Email：nkai103@yahoo.com.tw
Publishing Date　March 2014
Cover Story　Chengdu BAOLI Crown Plaza Hotel (Source: EDSA Orient)

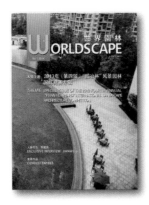

WORLDSCAPE 目录

世界园林 第五辑 2014年 第一期
主　题 2013年（第四届）"园冶杯"风景园林国际竞赛专辑

WORLDSCAPE
Vol 5 No.1 2014
THEME: SPECIAL ISSUE OF THE 2013 FOURTH ANNUAL "YUAN YE AWARD" INTERNATIONAL LANDSCAPE ARCHITECTURE COMPETITION

总编心语	014	
资讯	016	
	020	第四届园冶高峰论坛在北京隆重举行
	028	首届亚洲女风景园林师论坛暨亚洲园林协会女风景园林师分会成立大会成功举办
	034	"园冶杯"风景园林（毕业作品、论文）国际竞赛
	040	园冶访谈
	042	"园冶杯"获奖学生感言
人物专访	052	EDSA Orient 总裁 李建伟
竞赛作品	064	
一等奖	064	中国·云南·大山包国际重要湿地概念性景观规划设计 / 邵珊 李建磊 恭珉皓 孙计 张冬妮
	070	老公园更新演绎——以黄兴公园改造为例 / 荣南 陈荻
	076	城市绿洲，低碳生活——重庆南岸区兰花湖公园景观规划设计 / 张泽华
二等奖	086	以水为"纽带"链接城市间区域关系——盘锦滨海新区景观规划设计 / 贾晓丹
	090	衰落站区的新能量——南京浦口火车站及其周边地区景观重组与整合设计 / 蒋沂玲 段倩 徐佳琪
	096	土虱瓮·好挤好挤——台中北屯区军功寮人文景观调查规划 / 陈冠云 梁韦茵
	100	美丽乡村，生态家园——连云港市前云村改造规划设计 / 唐菲 吴安琪 李旭彤
	104	"追寻记忆里的蝉噪与鸟鸣"——上海嘉定北郊野公园湿地涵养林景区设计 / 黄川墅
	108	金三角绿洲——汕头市北郊公园规划设计 / 陈磊晶 郑懿 肖红涛
	112	微山微水微声——李光地读书处生态文化景观设计 / 侯喆昊
	116	针灸术——城市绿楔的后棕地生态修复 / 张姝 杨文祺
	120	铁路公社——南京浦口火车站改造设计分析 / 高枫 车建安 刘熠 王辉
设计类最佳人文奖	128	沈阳薰衣草庄园景观设计 / 陆书雯
	132	低碳、低术、低生活——农民工工地生活空间景观策略 / 高东东
三等奖	138	行走的花园模块——中关村二小校园景观设计 / 李杨
	142	武汉市黄陂区船舶主题儿童公园设计 / 刘悦 林黛
	146	曲水花洲——南宁青秀山水生花园概念设计 / 孙林琳
鼓励奖	152	马头山国家级自然保护区生态旅游区规划设计 / 杨天人 常凡 李菁
	156	田园城市——北京上地树村村有机更新设计 / 王艳秋 武键
	160	耕·记一城市绿色加工厂——重庆渝北区生态苗木产业园 / 余梅 李今朝 李月文
	164	LIGHT——城市公共空间小型构筑体概念设计 / 曾舜怡
	168	徘徊·循迹·新生——辽宁海城市教军山城市公园景观设计 / 马斯婷
	172	熊猫岛——雅安市熊猫岛主题公园景观规划与设计 / 周云婷
	176	沁园住宅区二期景观设计 / 吴竹韵
征稿启事	181	
广告索引	封二	"园冶杯"住宅景观奖竞赛
	封三	中国风景园林网
	002	"园冶杯"风景园林（毕业作品、论文）国际竞赛
	004	北京东方园林股份有限公司
	008	棕榈园林股份有限公司
	009	深圳市柏涛环境艺术设计有限公司
	010	阿拓拉斯规划设计有限公司
	013	北京夏岩园林文化艺术集团有限公司
	033	盛世绿源科技有限公司
	050	北京市园林古建设计研究院有限公司
	082	上海亦境建筑景观有限公司
	083	济南市园林规划设计研究院
	084	无锡绿洲景观规划设计院
	124	重庆华宇园林股份有限公司设计分公司
	125	安道国际
	126	源树景观
	136	天开园林景观工程有限公司
	150	杭州市园林绿化工程有限公司
	151	盛世绿源科技有限公司
	180	广州山水比德景观设计有限公司
	182	浙江青草地园林市政建设发展有限公司
	184	北京海韵天成景观规划设计有限公司

WORLDSCAPE

CONTENTS

EDITORIAL	014	
NEWS	016	
	020	THE 4TH YUAN YE FORUM WAS SUCCESSFULLY HELD IN BEIJING
	028	THE FIRST ASIAN WOMEN LANDSCAPE ARCHITECTS FORUM AND THE INAUGURAL MEETING FOR THE WOMEN LANDSCAPE ARCHITECTS BRANCH OF ASIAN LANDSCAPE ARCHITECTURE SOCIETY WERE SUCCESSFULLY HELD
	034	"YUAN YE AWARD" INTERNATIONAL LANDSCAPE ARCHITECTURE GRADUATION PROJECT/THESIS COMPETITION
	040	INTERVIEW OF "YUAN YE AWARD" COMPETITION
	042	PRIZE-WINNINGSTUDENTS' WORDS OF "YUAN YE AWARD" COMPETITION
EXCLUSIVE INTERVIEW	052	JIANWEI LI, THE PRESIDENT OF EDSA ORIENT
CONTEST ENTRIES	064	
THE FIRST PRIZE	064	DASHANBAO INTERNATIONAL WETLAND CONCEPTUAL LANDSCAPE PLANNING DESIGN IN YUNNAN, CHINA / Shan Shao Jianlei Li Minhao Gong Ji Sun Dongni Zhang
	070	THE PROGRESS OF UPDATING AN OLD PARK —— TRANSFORMATION OF HUANGXING PARK, SHANGHAI / Nan Rong Di Chen
	076	URBAN OASIS, LOW-CARBON LIFE —— LANDSCAPE DESIGN OF ORCHID LAKE PARK IN NANAN DISTRICT, CHONGQING / Zehua Zhang
THE SECOND PRIZE	086	"CITY BOND"AMONG REGIONAL RELATIONSHIPS——LANDSCAPE PLANNING OF PANJIN COASTAL NEW AREA / XiaodanJia
	090	REGENERATION OF THE DISUSED STATION——LANDSCAPE REGENERATION OF NANJING PUKOU RAILWAY STATION AND THE SURROUNDING AREAS /Yiling Jiang Qian Duan Jiaqi Xu
	096	TU SHR WENG · SIDE BY SIDE——LANDSCAPING CONSERVATION PROJECT FOR THE CATFISH HUMANE COMMUNITY/ Guanyun Chen Weiyin Liang
	100	BEAUTIFUL COUNTRYSIDE, ECOLOGICAL HOMELAND——PLANNING AND DESIGN OF QIANYUN VILLAGE IN LIANYUNGANG/ Fei Tang Anqi Wu Xutong Li
	104	"PURSUING THE SONG OF CICADA AND BIRDS IN THE MEMORY"——THE DESIGN OF WETLAND CONSERVATION FOREST IN NORTHERN JIADING / Chuanhe Huang
	108	THE GOLDEN TRIANGLE OASIS——NORTHERN SUBURB PARK PLANNING AND DESIGN IN SHANTOU / Leijing Chen Yi Zheng Hongtao Xiao
	112	MICRO HILL, MICRO WATER, MICRO SOUND——THE ECOLOGICAL CULTURAL LANDSCAPE DESIGN OF LIGUANGDI'S READING PLACE /Zhehao Hou
	116	ACUPUNCTURE——THE ECO-REMEDIATION OF POST-BROWNFILEDS IN A WEDGE-SHAPED GREEN SPACE / Shu Zhang Wenqi Yang
	120	RAILWAY COMMUNE——DESIGN AND ANALYSIS OF NANJING PUKOU RAILWAY STATION / Feng Gao Jian'an Che Yi Liu Hui Wang
THE BEST HUMANIST PRIZE	128	LAVENDER MANOR LANDSCAPE DESIGN, SHENYANG / Shuwen Lu
	132	LOW CARBON, LOW TECHNOLOGY, LOW IMPACT——MIGRANT WORKER RESIDENTIAL LANDSCAPE STRATEGY / Dongdong Gao
THE THIRD PRIZE	138	WALKING GARDEN MODULE——ZHONGGUANCUN SECOND PRIMARY SCHOOL CAMPUS LANDSCAPE DESIGN / Yang Li
	142	DESIGN OF THE SHIP THEMED CHILDREN'S PARK IN HUANGPI DISTRICT, WUHAN /Yue Liu Dai Lin
	146	BENDING WATER AND FLORAL SANDBAR——NANNING·QINGXIU MOUTAIN HYDROPHYTE GARDEN CONCEPTUAL DESIGN / Linlin Sun
THE HORORABLE MENTION	152	PLANNING OF MATOUSHAN NATIONAL NATURE RESERVE ECOTOURISM ZONE / Tianren Yang Fan Chang Jing Li
	156	ORGANIC GARDEN—— BEIJING SHANGDI TREE VILLAGE RENEWAL DESIGN / Yanqiu Wang Jian Wu
	160	URBAN GREEN PLANTS——ECOLOGICAL INDUSTRIAL PARK OF SEEDLINGS IN CHONGQING YUBEI DISTRICT / Mei Yu Jinzhao Li Yuewen Li
	164	LIGHT——CONCEPTUAL DESIGN OF SMALL STRUCTURES IN URBAN PUBLIC SPACES / Shunyi Zeng
	168	TRACKING·WANDERING·NEWBORN——LANDSCAPE DESIGN OF HAICHENG JIAOJUN MOUTAIN URBAN PARK IN LIAONING PROVINCE / Siting Ma
	172	PANDA ISLAND——LANDSCAPE PLANNING AND DESIGN OF PANDA ISLAND THEME PARK IN YAAN / Yunting Zhou
	176	LANDSCAPE DESIGN OF THE SECOND STAGE OF QINYUAN RESIDENTIAL AREA / Zhuyun Wu
	181	**Notes to Worldscape Contributors**

黄山国际中心

黄山国际中心,该项目荣获2013年第八届金盘奖全国最佳商业楼盘。项目位于安徽省黄山市屯溪区东南角黎阳镇内,黎阳古镇始建于东汉建安年间,距今1800年整,1000多米长的黎阳古街颇具韵味。景观设计以黎阳老街改造为核心,以符合现代生活方式的休闲商业广场为配套,突出徽州本地历史文化传承,并赋予其新的形象语言。

创新是设计之魂,质量是立足之本,服务是发展之道
INNOVATION IS THE SOUL OF DESIGN, QUALITY IS THE BESE, SERVICE IS THE WAY OF DEVELOPMENT

深圳市柏涛环境艺术设计有限公司　　BOTAO LANDSCAPE ART & DESIGN CO.,LTD
地址:深圳市南山区华侨城生态广场C栋101　　电话:0755 26919715　26913889　　网址:www.botaoead.com　　新浪微博:柏涛环境

阿拓拉斯十周年

中国　电话：0086
北京市海淀区知春路

常务理事单位

无锡绿洲景观规划设计院有限公司
EDSA Orient
北京大元盛泰景观规划设计研究有限公司
北京夏岩园林文化艺术集团有限公司
棕榈园林股份有限公司
岭南园林股份有限公司
北京源树景观规划设计事务所
北京欧亚联合城市规划设计院
重庆金点园林股份有限公司
重庆天开园林景观工程有限公司
杭州天香园林有限公司
北京市园林古建设计研究院有限公司
盛世绿源科技有限公司

理事单位

北京东方园林股份有限公司
重庆华宇园林股份有限公司设计分公司
枫彩农业科技有限公司
江苏山水建设集团有限公司
苏州新城园林发展有限公司
广东四季景山园林建设有限公司
北京乾景园林股份有限公司
广州市林华园林建设工程有限公司
广州山水比德景观设计有限公司

总编心语 / EDITORIAL

王小璘
Xiaolin Wang

　　为提高风景园林专业学生的创新意识及园林设计水平，加强国际间各相关院校的景观学术和专业交流，同时为毕业生们提供一个全面展示才华和互相学习交流的平台，2010年起由20余所国内外景观相关院校联合发起"园冶杯"风景园林国际竞赛，每年针对本科生和研究生的毕业论文和作品进行选拔。竞赛设置设计、规划、论文及园林植物研究等四类，邀请来自世界近百所院校的专家、学者对提交的作品（论文）选出一、二、三等奖、单项奖及鼓励奖的获奖作品，并进行全球巡展。至今百余所院校的八千余名师生参与了该竞赛，近千份优秀作品脱颖而出。

　　本期介绍2013年（第四届）"园冶杯"风景园林国际竞赛的规划及设计类获奖代表作品，以及对获奖学生的访谈和他们对大赛的寄语。在此，感谢各院校的积极参与、各位老师的辛勤指导、评审专家的精心评选和社会各界的大力赞助与支持！

　　本刊很荣幸邀请到李建伟先生进行专访。李先生是EDSA Orient总裁兼首席设计师，有着丰富的跨国设计及教育培训的经历。李先生的设计理念，展现其对本土文化和生态环境的尊重，以及对教育和景观实务应具备统筹能力的关注，值得园林教育和规划设计工作者参考借镜。他并以「机会总是给有准备的人」的亲身经验勉励年青人要不断学习，求得知识。

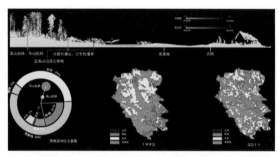

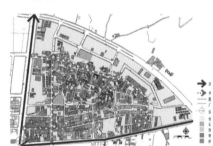
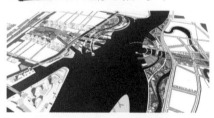
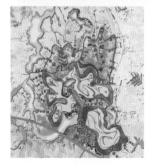

The "*Yuan Ye Award*" International Landscape Architecture Competition for undergraduate and graduates thesis and design works was established in 2010 as a joint initiative organized by 20 domestic and international landscape institutions. The goals of the initiative are to improve landscape students' innovative consciousness and design level; strengthen international academic and professional exchanges between landscape institutions; and provide a comprehensive display of talent, as well as a learning platform for mutual exchange between graduate students. The competition is divided into four categories: design, planning, research, and garden plants studies. Experts and scholars from hundreds of institutions around the world are invited annually for the selection of first, second, third, special mention and encouragement award prize winners. An international exhibition tour is held thereafter. Up to now, there has been more than eight thousand teachers and students from more than one hundred colleges participating in the contest, with nearly a thousand works selected as outstanding.

This issue introduces the winning works of the Planning and Design Category of The 2013 Fourth Annual "Yuan Ye Award" International Landscape Architecture Competition, along with interviews with the winning students. We would like to take the opportunity to thank the institutions for their active participation, the teachers for their guidance, and the experts for their careful selection and evaluation. We would also like to acknowledge the strong sponsorship and support of local communities and the population at large.

In this issue we have the honor to present an interview with Mr. Jianwei Li. Mr. Li is the EDSA Orient president and chief designer. He has an international experience in landscape design and education. His design philosophy promotes a heightened respect for domestic culture and ecological issues, and pay attention to the landscape management ability which can serve as a reference for both landscape educators and landscape designers. Mr. Li also encourages the younger generation to acquire a broad range of knowledge and skills, based on his experience that 'opportunities are always for those who are prepared'.

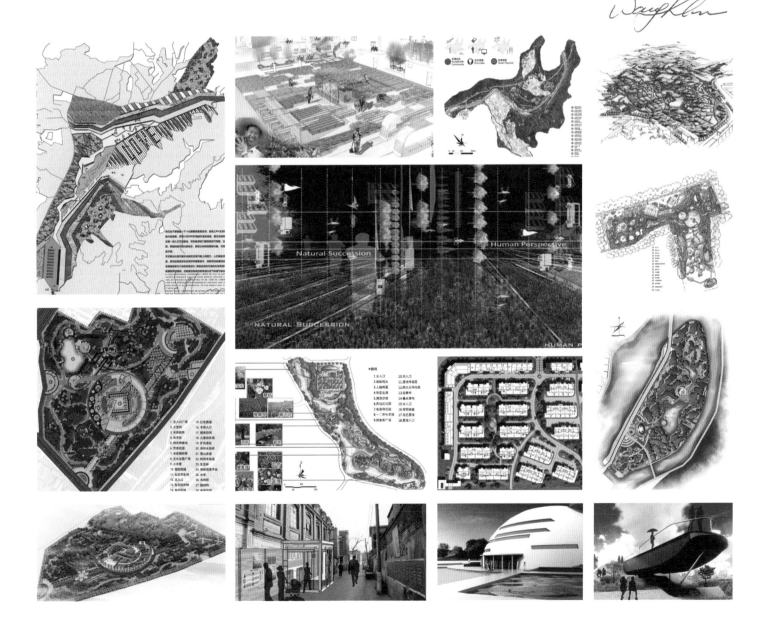

资讯 /NEWS

Kromhout 军营景观

Kromhout 军营所在的城市发展规划是由荷兰 Karres en Brands 景观设计事务所与 Van Schooten 一起制定的。他们的基本理念就是创建一个清晰和有组织的城市布局，同时保持一个美丽的外在形象。

场地分为三个区：一区内有一系列的办公大楼运行，以垂直于主要的道路的方式排列。引入护城河的设计理念，象征城市的防御形式。不同的建筑中间一共分布了七个花园和一个阅兵场。二区则旨在创建一个与相邻的大学相呼应的空间结构。该场地主要由石板道路、大草坪和大树组成景观框架。三区的中央公园则起到城市的绿色核心的作用，帮助维持复杂的生态平衡。

来源：筑龙网

阿兰·普罗沃作品：PRIEURE 城市开发区

这个充满活力的地带位于欧洲迪士尼公园入口处，实质上服务于第三产业。Signes 集团负责设计这个未来公园的所有公共区域和景观空间。规划方案从绿色轴线主题演变而来，设计构想结合了古典园林的元素，同时也具有英式园林的风格。

来源：中国风景园林网

波兰 Ruch Chorzów 体育场设计方案确定

由 GMT Myslowice 事务所完成的 Ruch Chorzów 体育场是波兰最大的足球体育场，该体育馆可以容纳 12,000 个座位，尽管方案被描述为"整体设计并非最出色"的设计，但它仍以简约、干净的外形和"物超所值"从多个方案中脱颖而出。Ruch Chorzów 体育场名字中的的"Ruch"在波兰语中有"运动"之意。作为呼应，设计理念围绕"当行人行走在体育场中可以移步换景，在不同角度体验到不一样的建筑之美。"

体育馆建成后，单层分布的看台将形成完整的碗状，可以容纳 12,000 个座位；另外，有关方面人士也设想过将其打造成可扩展的结构——在不改变设计完整性前提下，通过在顶部加三行以达到 4,000 个座位的需求量。该项目预计 2014 年中旬开始建设。

来源：Archdaily

大阪哈利·波特主题公园于 2014 年年底开园

位于大阪的"哈利波特魔法大世界"将于 2014 年年底正式开园迎客，造价预计将达 4.82 亿美元，引来无数粉丝翘首期待。

在"哈利·波特主题公园"里，你可以看到书中最具代表性的几处地点：包括霍格莫德小镇、神秘的禁林、霍格沃茨魔法学校、九又四分之三站台等多个地方。还可以深入到城堡中心，参观校长邓不利多的办公室，以及小镇中贩卖魔法用品的小店；可以在破釜酒吧里吃饭，在霍格莫德村的商店里购买巧克力蛙、黄油啤酒，在对角巷购买魔杖。不过恐怕仍然得用"麻瓜"世界里的货币——美元购买门票和上述商品。如果你觉得这还不够刺激，那么你可以体验在晚上自己选择住在霍格沃兹四个学院的休息室中的任意一个。

来源：留学生杂志

荷兰花园式学校大门

曾于 1959 年修建的阿尔玛花园式学校被阿姆斯特尔花园式学校所取代，在这里有超过 500 名小学生学习着自然科学。学校的两个标志性的大门均是由荷兰 Tjep 工作室所设计的，将许多当代的设计元素融入进了这片历史古迹中，用自然元素将大门点缀得栩栩如生。

主设计师 Frank Tjepkema 评论该作品："这耐人寻味的大门作品一方面表达了对花园式学校的敬意，另一方面也为校园带来了崭新的感官"。他还补充道："大门的底段伴有清晰可见的编织纹，门高4米，宽8米，是阿姆斯特尔城市到花园式学校的必经之地。我还希望来这里学习的孩子们，可以领略到大自然的奥秘之处"。

来源：筑龙网

荷兰水防线纪念园

该项目始建于 1794 年，早期是荷兰的一处军事堡垒，现已成为一处国家级的纪念区。Culemborg 市政府和 Werk aan't Spoel 基金会希望这处废弃的堡垒能重新成为引人注目的休闲胜地。经过改造，场地内新建了一座山庄，用于举办一系列由市民发起的活动。设计师将这些计划融入到了全方位的综合性设计之中，设计内容还包括昔日的水闸。

设计的灵感来自于场地内丰富的历史信息，可以将该灵感理解为一个巨大的草坪雕塑场地，结合了新旧历史元素，如战壕掩体、防弹构筑物和一个露天剧场。该项目将一些本土的地方性活动组织起来，因此形成了一种新型的公共空间，并且有潜力成为"新荷兰水线"上最具吸引力的景观之一。

来源：筑龙网

美国西雅图亚马逊公司总部新园区提案

这是由 NBBJ 设计的美国西雅图亚马逊公司总部新园区提案。该规划设计中，办公区中心建设3座自然生态博物馆。这三个植物园是三个圆顶结构，如同3个水泡。具有"轻视觉、有机几何形态和雕塑性"。根据最新的设计方案，新的计划将会增加公众参与的空间，其中1672平米用于零售业，还增加了一个 1.5 米宽的轨道，连接 Blanchard 街和第七大道。

全部规划共占地西雅图市中心丹尼三角区的3个街区，新的总部大楼包括3座38层的办公楼，2座中层办公楼和一座多功能会议中心，还包括公共公园、自然生态博物馆和零售空间。

来源：互联网

日本江岛大桥因坡度陡峭出名 前进百米升高约6米

连结日本鸟取县境港市和松江市的江岛大桥近日成为一个旅游新热点。

江岛大桥全长约1446米，高约44米，桥下可供 5000 吨级的轮船通过。松江市一侧的斜率为 6.1%，每前进 100 米升高约6米。从境港市一侧上桥，从"桥顶"眺望能够将中海湖的景色尽收眼底，黄昏时的美景也很著名。

如果踩油门过大则可能超速，当地呼吁人们安全驾驶，相关负责人表示："希望大家慢慢行驶，欣赏沿途风景。"

来源：环球网

首批基因改造自发光植物开始拍卖

Bioglow 公司的研究者将一种生物发光性细菌的基因，通过基因工程技术接入到一种普通的装饰性植物的花烟草上，从而创造出第一个家用生物光源，并被命名为星光阿凡达。这是第一种自发光植物，也是合成生物学向改造我们生活迈出地第一步。

目前正在拍卖第一批发光植物，他们在今年的晚些时候将会为订购者培育植物。尽管这些植物并不会发出太强的光，不过这释放了一个信号：在不远的将来将会有越来越多的基因改造植物来让我们的居住空间变得更加舒适和漂亮。

来源：Cnbeta 网站

希腊雅典城中心复兴竞赛获胜方案

这是由 OKRA 团队与瓦格宁根大学合作规划设计的雅典城市景观，使其成为当代大都市中心，并在此次国际竞赛中脱颖而出。

规划中尽量减少机动车辆的交通。"弹性城市"策略包括降低城市热度、改善热舒适性的具体态度，减少空气污染，减少能源使用和解决污水问题。将城市的三角形转化成绿色框架，来缓解城市热。绿化战略与水战略相结合，收集雨水用于灌溉。

设计还考虑到"无障碍城市"，恢复行走体验的连续性和恢复公众交通。同时该设计力图打造一个充满活力的"林荫大道"，加入小型的活动中心，从而建立环境建设和公共领域之间的联系以提高大众活动的积极性，为提高市民生活质量做出贡献。

来源：园林景观网

智利的山间庇护小屋

智利的建筑专业学生 Rodrigo Cáceres Céspedes 设计了一个半透明的庇护小屋，用来庇护来此旅行的自行车骑行者。整个装置由轻质铝结构和双层的特殊布料结合完成，整个设计使得来安第斯山脉的游客更加惬意，设计同时也考虑了可持续性，以延长整个结构的使用寿命。

整个设计的思想是无限循环，可以用于自行车训练和比赛，当然也可以作为旅游用途，整个庇护小屋有可调节的支撑结构来适应不规则的地形，而且整个小屋比较容易搭建和移动。脚踏式发电机和太阳能热水器为自行车骑行者提供了必要的帮助，在这个独特的空间里，大家不仅可以享受灿烂的风景，也可以分享旅途的故事。

来源：筑龙网

2013ASLA 专业奖：布鲁克林植物园游客中心

布鲁克林植物园游客中心荣获 2013ASLA 专业奖、通用设计荣誉奖。该设计将建筑、绿色屋顶和 3 英亩的植物园游客中心的景观环境整合起来，作为通向布鲁克林植物园的新通道。建筑采用了曲折的形式作为对植物园内山势的一种延伸，景观设计与建筑结构完美融为一体。这一景观系统包括：绿色屋顶，生态洼地，环保渗透地面以及富有特色的雨水花园。其核心特征就是建筑物上 10000 平方英尺的屋顶绿化。这个可以容纳超过 40000 株植物的屋顶花园，使植物的根系能在屋顶的负荷限度内最大化地深度生长。通过将植物与建筑融为一体，绿色屋顶重新定义了游客与花园、展览与运动、文化与环境之间物理和哲学上的关系。护堤与池塘之间的关联互动，用阴阳相生的方法进一步阐释了这种将建筑与景观环境相融合的独特并且可持续的设计途径。

来源：筑龙网

2014 美国建筑师学会城市规划荣誉大奖公布

美国建筑师学会评出了 6 个项目授予城市规划奖，来表彰他们在建筑、城市与地区规划和社区建设方面的突出贡献。AIA 将在 2014 年全国大会暨设计博览会上为他们颁奖。

1）阿肯萨斯大学社区规划中心和 Marlon Blackwell 共同设计的小石城城市走廊。2）Skidmore, Owings 和 Merrill LLP 对丹佛联合车站社区的重建。3）WXY architecture + urban design 设计的曼哈顿东河滨水"蓝道"。4）Duany Plater-Zyberk 和 Co. LLC 的迈阿密 21 区改造。5）Lake Flato Architects 的珍珠酿酒厂促进经济总规划。6）Skidmore, Owings 和 Merrill LLP 的山茶半岛战略规划。

来源：互联网、中国风景园林网

2015年米兰世博会伊朗馆方案

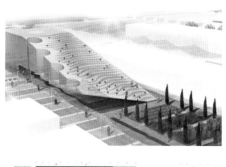

Laboratorio di Architettura e Design（LAD）与Naqsh, E, Jahan-Pars（NJP）合作设计的伊朗馆方案，被选为2015米兰世博会的伊朗馆国际比赛方案的冠军。以"伊朗高原上的生命历程"为核心概念，发扬了伊朗用地下暗渠将水引到沙漠地区的理念，这个获奖作品完全符合米兰世博会"喂养地球"这一主题。该作品重点将两种现代建筑元素运用到了设计中：花园和"暗渠"。水

通过挖好的地下管道从高原输送到城市。展馆引领游客在"暗渠"中通过，最终到达得到灌溉的田地。而花园也会种植一些作物，以便游客能在世博会品尝伊朗美食。虽然该作品赢得了竞赛，但是伊朗政府方面并没有表明是否要采纳此项设计。

来源：筑龙网

迪勒和伦弗罗赢得莫斯科公园竞争

美国的迪勒—斯科菲迪奥—伦弗罗建筑事务所（Diller Scofidio & Renfro）赢得了一场国际竞争：在莫斯科市中心设计一个新的公园。"查尔亚德耶"（Zaryadye）在莫斯科河边，靠近克里姆林宫和红场，是莫斯科中央的一个历史性区域。这个新的面积为130,000平方米的公园，将建设在原"俄罗斯酒店"（Hotel Russia）的位置上。这个获胜的初步规划，包括4个典型的俄罗斯景观——苔原、干旱草原、森林和沼泽地。这4个景观从高到低分布在一片台地上，从东北向西南相互交叉。该公园的建设工作预期在明年年底开始，在2016年完工。

来源：筑龙网

俄罗斯中央公园国际设计竞赛方案

这是荷兰HOSPER参加俄罗斯公园国际设计竞赛的方案，在最后的角逐中，获得五个入围团队中的第二名。基地内有一座30米高的小山，内部现存的的森林和水系作为此次公园设计和布局的基础。公园被分为六个功能单元，每一个单元都有一个主题，建筑形式相应地也对这些主题进行了呼应。每个单元中的建筑都是不同的，但是也有一些共同点：建筑顶部有一个绿色空间和娱乐路径，每组建筑都有一个很大的玻璃圆顶。公园中标志性的金色球体，不仅标示了这里是公园的主入口，也将成为公园的交通枢纽。

来源：中国风景园林网

线立方景观

"线立方"是参加2013年中国国际建筑艺术双年展的作品之一。整个构筑物约16米高，像一个三维编织的挂毯，整体架构全部由金属框架支撑，而且提供了与众不同的感官体验，整个结构从各个角度观赏有不同的变化。

设计源于6个简单框架的扭曲与组合，每个框架中都包含一个二维组合制成的绳子，通过建立横向轴、类似编织品的连接方式来增加整个结构的深度和空间，并在某种程度上突破了多维数据，建立了一些不稳定的角。

结构设计师试图创造一个演进的经验效应——从外部理解一个几何对象、一种外壳、一个动态的变化轨迹。

来源：筑龙网

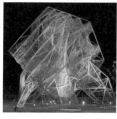 

第四届园冶高峰论坛在北京隆重举行
THE 4TH YUAN YE FORUM WAS SUCCESSFULLY HELD IN BEIJING

住房与城乡建设部副部长仇保兴
特为大会发来贺信

2014年1月9日–11日第四届园冶高峰论坛在北京新大都饭店成功举办，论坛主题为"促进生态文明，发展美丽中国"。作为风景园林行业重要的年度盛会，园冶高峰论坛已经成功举办了三届。

本届论坛的开幕式由住房和城乡建设部城建司原副巡视员、中国风景名胜区协会副会长曹南燕主持。住建部副部长仇保兴为此届论坛发来贺信。受邀嘉宾有：中国工程院院士孟兆祯，中国工程院院士卢耀如，中国科学院院士傅伯杰，世界风景园林大师、国际绿色建筑与住宅景观协会副主席亨利·巴瓦，亚洲园林协会主席、台湾造园景观学会名誉理事长、《世界园林》杂志总编王小璘，住建部原总经济师、住建部科技委副主任、中国建筑装饰协会会长李秉仁，北京市园林绿化局副局长强健，韩国生态景观协会会长、韩国江陵大学环境造景学科教授赵泰东等人。同时出席开幕式的还包括环境保护部、国土资源部、国家林业局以及各省市政府园林局、农业局、园林绿化协会、园艺学会的领导，国内外专家学者，国内外知名企业负责人和业界同行等共计700余人。

开幕式上举行了"2013'园冶杯'风景园林（毕业设计、论文）国际竞赛""园冶杯住宅景观奖""城市园林绿化综合竞争力百强暨十佳园林企业"等颁奖典礼。

弘扬园冶造园艺术，发展风景园林事业

住房和城乡建设部副部长　仇保兴

各位专家学者，同志们：

在全国各地全面贯彻落实党的十八届三中全会和中央城镇化工作会议精神之际，召开第四届园冶高峰论坛，很及时也非常有意义，首先我代表住建部向大会的召开表示热烈的祝贺！向来自世界各国的风景园林师们表示热烈的欢迎！向园冶杯风景园林国际竞赛获奖的师生、单位表示祝贺！

党的十八届三中全会明确提出了紧紧围绕建设美丽中国，加快建立生态文明制度，建立国家公园体制，形成人与自然和谐发展现代化建设新格局。刚刚结束的中央城镇化工作会议，党中央、国务院提出要紧紧围绕提高城镇化发展质量，高度重视生态安全，扩大湖泊、湿地等绿色生态空间比重，优化城镇化布局和形态。发展有历史记忆、地域特色、民族特点的美丽城镇。把城市放在大自然中，把绿水青山保留给城市居民。这是在新时期新形势下，党中央、国务院做出的重大决策，这给我们城市建设和风景园林工作者带来个新的历史机遇，同时也面临着极大挑战。

推进新兴城镇化，需要我们广大设计师的智慧和创新，加强生态文明，建设美丽家园成为当代风景园林师的神圣责任、历史使命。希望通过园冶高峰论坛，搭建国际交流合作的平台，汇集国内外的专家学者，积极的探索发展，努力实现共同繁荣。

最后，在新春佳节即将到来之际，祝各位专家学者、行业同仁新年快乐，万事如意！

仇保兴

二〇一四年一月六日

中国工程院院士 孟兆祯　　中国工程院院士 卢耀如　　中国科学院院士 傅伯杰　　世界风景园林大师 亨利·巴瓦　　亚洲园林协会主席、台湾造园景观学会名誉理事长、《世界园林》杂志总编 王小璘　　住建部原总经济师、住建部科技委副主任、中国建筑装饰协会会长 李秉仁

论坛会场

中国城市园林绿化百强企业颁奖　　园林绿化行业十佳企业颁奖　　国际住宅景观奖颁奖　　"园冶杯"风景园林国际竞赛颁奖

颁奖典礼后，孟兆祯、卢耀如、傅伯杰等院士，先后就中国风景绿化园林行业现状、政策方向、发展趋势、发展对策等方面做了精彩的报告分析。

9日晚上的世界风景园林大师对话活动邀请到了亚洲园林协会主席、台湾造园景观学会名誉理事长、《世界园林》杂志主编王小璘，世界风景园林大师、国际绿色建筑与住宅景观协会副主席亨利·巴瓦，著名景观设计师、Studio Outside 合伙人萨尔瓦多，清华大学景观系副主任朱育帆，北京林业大学教授朱建宁，华南农大城市规划与园林系教授李敏。本活动由亚洲园林协会主席、台湾造园景观学会名誉理事长、《世界园林》杂志主编王小璘教授主持。

10日–11日举行了园林城市与人居环境高峰论坛、生态城市与城市设计学术研讨会、欧亚设计师学术交流会、休闲度假与住宅景观设计交流会、女风景园林师座谈会、园林植物与工程设计研讨会、风景园林高校论坛，以及园林工程技术与企业管理研讨会等8个分论坛。

专家院士合影1

专家院士合影2

与会嘉宾合影

中国建设教育协会会长 李竹成

分论坛会场1

分论坛会场2

分论坛之一 —— 园林城市与人居环境高峰论坛

主持嘉宾：住建部城建司原副巡视员、中国风景名胜区协会副会长 曹南燕

主持嘉宾：湖南建设厅城建处原处长 周淑兰

北京市园林绿化局副局长 强健

北京林业大学教授 朱建宁

台湾东海大学教授 方伟达

北京市公园管理中心总工程师 李炜民

乌鲁木齐园林管理局局长 翟勤盈

该分论坛由住建部城建司原副巡视员、中国风景名胜区协会副会长曹南燕及湖南建设厅城建处原处长周淑兰主持，众多专家学者及景观设计企业代表出席。会上，北京市园林绿化局副局长强健、北京林业大学教授朱建宁、台湾东海大学教授方伟达、北京市公园管理中心总工程师李炜民、深圳市北林苑景观及建筑规划设计院有限公司院长何昉、乌鲁木齐园林管理局局长翟勤盈等分别从不同方面讲述了园林绿化行业的发展，分享了自己的实践经验，覆盖范围由内地到港澳台地区，横贯中西。

分论坛之二 —— 欧亚设计师学术交流会

主持嘉宾：南京林业大学风景园林学院副院长 张青萍 | 华南农业大学城市风景园林与城市规划系教授 李敏 | 清华大学景观系副主任、教授 朱育帆 | 亚洲园林协会副会长、韩国生态景观协会会长、韩国江陵国立大学教授 赵泰东 | 韩国著名景观设计师 安翼东 | 沈阳建筑大学教授 陈雪松

欧亚设计师学术交流会会场

该分论坛由南京林业大学园林学院副院长张青萍主持。华南农业大学风景园林与城市规划系教授李敏，清华大学景观系副主任、教授朱育帆，亚洲园林协会副会长、韩国生态景观协会会长、韩国江陵国立大学教授赵泰东，韩国的著名设计师安翼东，沈阳建筑大学教授陈雪松等发表了精彩演讲。之后，还开辟了互动环节，台下的设计师、学生与媒体记者就汉江河道设计、冬季奥运会植物选择、废旧河道设计等问题进行了热烈的讨论。

分论坛之三 —— 生态城市与城市设计学术研讨会

中国风景园林学会原副理事长、北京市园林局原副局长 张树林 | 世界风景园林大师 亨利·巴瓦 | EDSA Orient 总裁兼首席设计师 李建伟 | Studio Outside 设计师 萨尔瓦多 | 北京源树景观设计所首席设计师 白祖华 | James Corner Field Operations 设计师 成行

该分论坛由中国风景园林学会原副理事长、北京市园林局原副局长张树林主持。世界风景园林大师亨利·巴瓦、EDSA Orient 总裁兼首席设计师李建伟、Studio Outside 设计师萨尔瓦多、北京源树景观设计所首席设计师白祖华、James Corner Field Operations 设计师成行、沈阳建筑大学建筑与规划学院副院长姚宏韬、南京赛诺格顿景观工程有限公司设计总监张增记出席了该研讨会，并就城市景观设计、城市规划等热点话题，结合实际案例做了演讲。强大的嘉宾阵容及精彩的演讲，吸引了很多内业人士及高校师生参加，整个会场座无虚席。

生态城市与城市设计学术研讨会会场

分论坛之四 —— 休闲度假与住宅景观设计交流会

主持嘉宾：北京市园林绿化局绿化处处长 崔勇　　龙湖集团景观中心负责人 何格　　北京源树景观设计事务所合伙人、总经理、首席设计师 白祖华　　华润置地（北京大区）景观设计总监 姚鹏

重庆天开园林副总裁 俎志峰　　广州山水比德景观设计有限公司副总经理兼设计分公司设计总监 利征　　中外园林集团设计院院长 孟欣　　北京市园林古建院YWA工作室首席设计师 严伟

该分论坛由北京市园林绿化局处长崔勇主持。龙湖集团景观中心负责人何格，北京源树景观设计事务所合伙人、总经理、首席设计师白祖华，华润置地（北京大区）景观设计总监姚鹏，重庆天开园林副总裁俎志峰，广州山水比德景观设计有限公司副总经理兼设计分公司设计总监利征，中外园林集团设计院院长孟欣，北京市园林古建院YWA工作室首席设计师严伟等，从甲方或乙方的角度对当今国内住宅景观设计及其未来如何更好地与休闲度假相结合进行了深入的交流和探讨。

休闲度假与住宅景观设计交流会会场

女风景园林师合影

分论坛之五 —— 女风景园林师座谈会

主持嘉宾 住建部城建司原副巡视员、中国风景名胜区协会副会长 曹南燕

该座谈会由中国风景名胜区协会副会长曹南燕主持。来自风景园林界各企事业单位的女风景园林师以及各高校师生女性代表们四十余人共聚一堂,成为本届园冶高峰论坛的一道亮丽风景线。住建部城建司巡视员陈蓁蓁、东方园林董事长何巧女、北京市园林局原副局长张树林、北京林业大学园林学院教授杨赉丽、北京市公园管理中心巡视员刘英、湖南省住建厅城建处原处长周淑兰、宁夏建设厅城建处原处长陈维华、山西省风景名胜监管中心副主任牛峥嵘、河北风景园林协会常务副理事长王景、四川风景园林协会副会长肖晴等,结合各自在实际工作中的经验,针对如何把中国风景园林做得更好各抒己见。

住建部城建司巡视员 陈蓁蓁

东方园林董事长 何巧女

中国风景园林学会原副理事长、北京市园林局原副局长 张树林

北京林业大学园林学院教授 杨赉丽

北京市公园管理中心巡视员 刘英

湖南省住建厅城建处原处长 周淑兰

宁夏建设厅城建处原处长 陈维华

山西省风景名胜监管中心副主任 牛峥嵘

河北风景园林协会常务副理事长 王景

四川风景园林协会副会长 肖晴

分论坛之六 —— 风景园林高校论坛

华南农业大学风景园林与城市规划系
教授 李敏

南京林业大学风景园林学院副院长 张青萍

中南林业科技大学风景园林学院副院长、教授 胡希军

四川农业大学园林学院院长 陈其兵

天津大学建筑学院环境艺术设计系教授 刘庭风

中国农业大学副教授 奚雪松

获奖学生：南京林业大学风景园林学院 陈荻

获奖学生：西南大学园艺园林学院 张泽华

获奖学生：苏州科技学院建筑与城市规划学院 唐瓴

该分论坛由华南农业大学风景园林与城市规划系教授李敏主持。南京林业大学风景园林学院副院长张青萍，中南林业科技大学教授胡希军，四川农业大学园林学院院长陈其兵，天津大学建筑学院环境艺术设计系教授刘庭风，中国农业大学副教授奚雪松等人做了精彩的演讲。之后，2013"园冶杯"风景园林国际竞赛获奖学生：包括南京林业大学风景园林学院陈荻、西南大学园艺园林学院张泽华、苏州科技学院建筑与城市规划学院唐瓴，向大家分享了各自获奖作品的设计思路与设计经验。

风景园林高校论坛会场现场

分论坛之七 —— 园林工程技术与企业管理研讨会

住建部城建司原副巡视员、中国风景名胜区协会副会长 曹南燕

陕西旅游集团公司少华山公司总经理、高级工程师 陈青

韩国景观设计师 徐尚日

济南园林局原局长 王玉华

该分论坛由住建部城建司原副巡视员、中国风景名胜区协会副会长曹南燕，陕西旅游集团公司少华山公司总经理、高级工程师陈青主持。韩国景观设计师徐尚日、济南园林局原局长王玉华，以及几十家园林企业代表，就园林绿化的设计、施工、养护、管理以及园林行业的发展趋势等多方面的问题进行了广泛深入的交流。

宁夏银川弘地绿化工程公司法人代表 李立

大千生态景观股份有限公司北京分公司总经理 窦阳

园林工程技术与企业管理研讨会会场

分论坛之八 —— 园林植物与工程设计研讨会

主持嘉宾：河北省风景园林与自然遗产管理中心副主任 孟欣

北京林业大学博士生导师 董丽

北京林业大学园林学院副教授 瞿志

该分论坛由河北省风景园林与自然遗产管理中心副主任孟欣主持。北京林业大学博士生导师董丽、北京林业大学园林学院副教授瞿志、河北农业大学风景园林学院院长黄大庄、北京林业大学园林学院副教授于晓南、中南林业科技大学讲师邢文等人，分别就园林植物的种植、养护、育种以及在风景园林中的应用做了精彩报告。

河北农业大学风景园林学院院长 黄大庄

北京林业大学园林学院副教授 于晓南

中南林业科技大学讲师 邢文

住房与城乡建设部副部长仇保兴
特为大会发来贺信

首届亚洲女风景园林师论坛暨亚洲园林协会女风景园林师分会成立大会成功举办

THE FIRST ASIAN WOMEN LANDSCAPE ARCHITECTS FORUM AND THE INAUGURAL MEETING FOR THE WOMEN LANDSCAPE ARCHITECTS BRANCH OF ASIAN LANDSCAPE ARCHITECTURE SOCIETY WERE SUCCESSFULLY HELD

弘扬亚洲造园艺术，发展风景园林事业

住房和城乡建设部副部长 仇保兴

尊敬的亚洲园林协会王小璘主席：

各位专家学者，各位女风景园林师：

在国际劳动妇女节之际，亚洲各国的女风景园林师汇聚一堂召开亚洲女风景园林师论坛，很及时也非常有意义，首先我代表住建部向大会的召开表示热烈的祝贺！向来自亚洲各国的女风景园林师们表示热烈的欢迎！

中国共产党十八届三中全会明确提出了加快建立生态文明制度，建立国家公园体制，形成人与自然和谐发展现代化建设新格局。刚刚结束的中央城镇化工作会议，提出要高度重视生态安全，把城市放在大自然中，把绿水青山保留给城市居民。这是在新时期新形势下，党中央、国务院做出的重大决策，这给我们城市建设和风景园林工作者带来个新的历史机遇，同时也面临着极大挑战。希望通过亚洲女风景园林师的平台，搭建国际交流合作的平台，发挥女风景园林师的智慧和力量，汇集国内外先进经验和创新方法，努力实现共同繁荣。同时也希望亚洲园林协会女风景园林分会积极团结各国的女风景园林师，发挥女性的聪明智慧，努力建成温馨的亚洲女风景园林师之家。

最后，祝各位女风景园林师妇女节快乐，万事如意！

二〇一四年三月三日

3月9日，首届亚洲女风景园林师论坛暨亚洲园林协会女风景园林师分会成立大会在北京新大都饭店成功举办。来自全国26个省、市、自治区、港台地区及日韩等多个亚洲国家的风景园林工作者百余人参加了大会。

大会由亚洲园林协会筹备委员会副主任、原北京市园林局副局长张树林主持。住房与城乡建设部副部长仇保兴特为大会发来贺信："对亚洲女风景园林师论坛的召开表示热烈的祝贺！向来自亚洲各国的女风景园林师们表示热烈的欢迎。希望通过亚洲女风景园林师的平台，搭建国际交流合作平台，发挥女风景园林师的智慧和力量，汇集国内外先进经验和创新方法，努力实现共同繁荣。同时也希望亚洲园林协会女风景园林分会积极团结各国女风景园林师，发挥女性的聪明智慧，努力建成温馨的亚洲女风景园林师之家"。

亚洲园林协会女风景园林师分会筹备委员会主任、中国风景名胜区协会副会长曹南燕对分会的筹备情况做了报告。筹备工作得到了住建部领导的高度重视，住建部副部长仇保兴对女风景园林师分会的筹备工作做了"团结女风景园林师，将中国大地设计建设的更美丽"的重要批示，使筹备工作得以顺利进行。建设部原副部长姚兵为大会题字"颂亚洲园林协会女风景园林师分会成立，弘扬繁荣风景园林学术作品"。我国著名古建保护专家，98岁高龄的郑孝燮先生亲笔为分会的成立题字"热烈祝贺亚洲园林协会女风景园林师分会成立"。

The first Asian Women Landscape Architects Forum and the inaugural meeting for the women landscape architects branch of Asian Landscape Architecture Society was held in Beijing Xindadu Hotel on March 9th. More than 100 landscape architects from Asian areas attended the meeting.

Baoxing Qiu, as vice minister of MOHURD (Ministry of Housing and Urban-Rural Development of People's Republic of China) made important instructions to the preparations of the women landscape architects branch and sent a congratulation letter for the meeting. Bing Yao (the former vice minister of Ministry of Construction) and Xiaoxie Zheng (China's famous expert in ancient architecture protection, 98 years old) inscribed for the establishment of the women landscape architects branch.

首届亚洲女风景园林师论坛会场

亚洲园林协会女风景园林师分会荣誉会长、第十及十一届全国政协委员、中国市长协会专职副会长、女市长分会执行会长 陶斯亮

亚洲园林协会女风景园林师分会荣誉会长、国务院参事、中国规划协会女规划师专业委员会主任、中国城市规划设计研究院顾问 王静霞

亚洲园林协会主席、《世界园林》杂志总编、台湾造园景观协会名誉理事长 王小璘

亚洲园林协会女风景园林师分会会长、中国风景名胜区协会副会长 曹南燕

中国城市规划协会副会长兼秘书长 王燕

亚洲园林协会女风景园林师分会副会长、北京林业大学教授 杨赉丽

香港园境师学会前会长、香港园境师注册局局长陈弘志代表女风景园林师之友致辞

亚洲园林协会女风景园林师分会副会长、原北京市园林局副局长 张树林

亚洲园林协会女风景园林师分会副会长 谭馨

日本国立千叶大学教授、造园景观协会常务理事 池边好

亚洲园林协会女风景园林师分会副会长 任春秀

北京公园管理中心巡视员 刘英

亚洲园林协会女风景园林师分会筹备委员会副主任、东方园林董事长何巧女宣读了女风景园林师分会理事、常务理事、会长、副会长、监事长、秘书长等候选人建议名单；聘请陶斯亮、王静霞为荣誉会长，王小璘为名誉会长，聘请谭庆琏、赵宝江、姚兵、孙筱翔、孟兆祯为高级顾问，聘请王迁、强健、陈弘志（香港）、王浩、王脩珺、赵泰东（韩国）、包满珠、李炜民等近30位同志为女风景园林师之友。

大会一致通过曹南燕当选女风景园林师分会会长，王秀娟（台湾）、任春秀、池边好（日本）、何巧女、杨赉丽、张树林、谭馨等任副会长，王玉华任监事长，陈青任秘书长。

中国城市规划协会副会长兼秘书长王燕，日本国立千叶大学教授、造园景观协会常务理事池边好女士，北京公园管理中心巡视员刘英，四川省风景园林协会副会长兼秘书长肖晴分别做了发言。亚洲园林协会女风景园林师分会荣誉会长、国务院参事、中国规划协会女规划师专业委员会主任、中国城市规划设计研究院顾问王静霞，亚洲园林协会女风景园林师分会荣誉会长、第十、十一届全国政协委员、中国市长协会专职副会长、女市长分会执行会长陶斯亮，亚洲园林协会主席、《世界园林》杂志总编、台湾造园景观协会名誉理事长王小璘分别为大会致辞。

亚洲园林协会女风景园林师分会成立大会

亚洲园林协会副主席、韩国生态造景协会会长、韩国江陵原州国立大学教授 赵泰东

亚洲园林协会女风景园林师分会监事长 王玉华

亚洲园林协会女风景园林师分会副会长、东方园林董事长 何巧女

台湾景观工程商业同业公会全国联合会理事长 湛锦源

亚洲园林协会女风景园林师分会名誉副会长 周淑兰

台湾景观学会副理事长、辅仁大学艺术学院景观系主任 王秀娟

韩国Inter造景技术师事务所董事长 金秀妍

四川省风景园林协会副会长兼秘书长 肖晴

中国城市规划协会、日本造园景观协会、中国风景名胜区协会、中国市长协会女市长分会、北京市公园管理中心、宁夏园林景观协会、武夷山风景名胜区管理委员会、西安市土木建筑学会风景园林专业委员会、山西省世界遗产和风景名胜区监管中心、中国会展经济研究会、河北省风景园林中心、吉林省园林协会、四川省风景园林协会等各地协会以及北京东方园林股份有限公司、北京易兰建筑规划设计有限公司、西安工业大学建工学院、武汉构城设计咨询有限公司、中国格申文化发展有限公司、宜时（北京）文化有限公司等企业和院校纷纷发来贺信。北京格申工艺美术制品有限公司还为亚洲园林协会女风景园林师分会赠送了具有中国传统文化特色的缶，祝贺分会成立。

首届亚洲女风景园林师论坛会场气氛活跃

首届亚洲女风景园林师论坛会场

女风景园林师分会揭牌仪式

赠缶仪式

具有中国传统文化特色的缶

亚洲园林协会主席王小璘、中国女市长协会会长陶斯亮、中国规划学会女规划师专业委员会主任王静霞为女风景园林师分会揭牌。北京格申工艺美术制品有限公司为亚洲园林协会女风景园林师分会赠送具有中国传统文化特色的缶。

Xiaolin Wang (the president of Asian landscape architecture Society), Siliang Tao (the president of China Female Mayors Association) and Jingxia Wang (the president of Female Planners Committee of China Association of City Planning) unveiled the plaque to open the Women Landscape Architects Branch. Beijing G&S Arts and Crafts Co., Ltd. presented a Fou with characteristics of Chinese traditional culture to the Women Landscape Architects Branch.

与会嘉宾1

与会嘉宾2

与会嘉宾3

与会嘉宾4

3月9日下午，举办了首届亚洲女风景园林师论坛。论坛由亚洲园林协会女风景园林师分会会长曹南燕主持。香港园境师学会前会长、香港园境师注册局局长陈弘志代表女风景园林师之友为论坛致辞。亚洲园林协会主席、《世界园林》杂志总编、台湾造园景观协会名誉理事长王小璘作了题为《面向逆境挑战的景观设计》的精彩报告。亚洲园林协会副主席、韩国生态造景协会会长、韩国江陵原州国立大学教授赵泰东，台湾景观工程商业同业公会全国联合会理事长湛锦源，分别就韩国园林行业发展现状、台湾园林行业发展现状以及自己的研究方向做了报告。香港园境师学会前会长、香港园境师注册局局长陈弘志，台湾景观学会副理事长、辅仁大学艺术学院景观系主任王秀娟，韩国Inter造景技术师事务所董事长金秀妍分别作了题为《香港风景园林发展的里程碑》《台北绿色城市创意行动》《园林设计的原则》的精彩讲座，与会嘉宾反响强烈，会场座无虚席。

首届亚洲女风景园林师论坛暨亚洲园林协会女风景园林师分会成立大会的成功举办，提升了女风景园林师在行业中的地位，切实推动了新形势下女风景园林师在国际上的交流与合作，积极推动着风景园林事业的健康蓬勃发展！

仇保兴副部长照片来源：http://www.mohurd.gov.cn/

"园冶杯"风景园林
（毕业作品、论文）国际竞赛

"YUAN YE AWARD" INTERNATIONAL LANDSCAPE
ARCHITECTURE GRADUATION PROJECT / THESIS COMPETITION

国际院校 积极参与
INTERNATIONAL INSTITUTIONS, ACTIVE PARTICIPATION

"园冶杯"风景园林(毕业作品、论文)国际竞赛(简称"园冶杯"竞赛)是每年在风景园林相关专业院校的毕业生中开展的毕业作品(论文)竞赛评选活动。其宗旨是促进风景园林行业的和谐发展,提高风景园林专业学生的创新能力及水平,展示学子风采,打造一个全面展示才华和互相学习交流的平台。同时进一步推动各相关院校和国际间的专业交流。

"园冶杯"竞赛已成功举办四届,自竞赛举办以来,国内外近百所相关院校、8000多名师生和业内专家、学者积极参与,产生了许多优秀的作品。

"Yuan Ye Award" International Landscape Architecture Graduation Project/Thesis Competition (short for "Yuan Ye Award" Competition)is a graduation project/thesis competition held among graduates of Landscape architecture-related colleges. Its aim is to promote the harmonious development of Landscape architecture, enhance the innovation capacity of students majoring in Landscape architecture, build a platform for fully displaying students' talents and mutually communicating, and boost professional communication among relative colleges and countries.

"Yuan Ye Award" Competition has been successfully held for four times, since it was held, 100 domestic and foreign universities , more than 8000 teachers and students, experts, scholars participated in the competition, many outstanding works were generated.

严格把关 有序评审
STRICT CONTROLLING, ORDERLY REVIEWING

"园冶杯"风景园林(毕业作品、论文)国际竞赛分为四类,分别是风景园林设计作品类、风景园林规划作品类、园林规划设计论文类、园林植物研究论文类。每类作品(论文)评选设置为两组:本科组和硕博组。

"园冶杯"风景园林国际竞赛每年由来自世界各地的150多位行业顶级专家教授组成评审组,秉承"公平、公正、公开"的评审原则,严格按照大赛规程,对各个参赛作品(论文)认真进行初审、复审、终审,以评选出最优秀的作品,选拔出最杰出的高校师生。

颁奖大会 隆重热烈
AWARDS CEREMONY, GRAND AND ENTHUSIASTIC

颁奖大会上，来自国内外各个高校的获奖师生欢聚一堂，共同分享设计理念与创作思路，表达获奖的喜悦与感激之情。百余所参赛院校的代表嘉宾，亲临盛会，与获奖师生互相交流。

高校论坛 学术盛宴
INSTITUTION FORUMS, ACADEMIC BOOM

高峰论坛中，全球风景园林行业知名专家学者汇聚一堂，交流最新的学术成果，探讨行业最新发展趋势。学术思想、百家争鸣；学术观点，异彩纷呈。

2010年首届园冶国际论坛在南京林业大学举行。南京林业大学园林学院党委书记王良桂主持会议。会上，共有7位来自各大院校的专家、教授进行了主题发言，并与在场师生进行了互动交流。

与首届论坛相比，第二届和第三届园冶国际论坛的组委会邀请了更多的国内外专家学者参与，同时举办了"风景园林师沙龙——对话大师"，"魅力中国——当代风景园林师的责任与义务"等数个多主题的分论坛。论坛为专家、学者们提供了一个学术交流机会，同时也让青年风景园林师充分领略了大师们的风采。

全球巡展 影响广泛
GLOBAL TOUR, EXTENSIVE INFLUENCE

"园冶杯"风景园林国际竞赛获奖作品在全国各地巡展，受到各大高校师生的关注。巡展期间，巡展单位与参观师生及设计师采取不同形式，从多个角度进行交流。"园冶杯"作品院校巡展在同学间产生了广泛的影响，增加了风景园林相关专业同学的学习素材。参观巡展的同学们都表示巡展的形式更有利于讨论和学习。

邵珊　　李建磊　　陈泳　　荣南　　张泽华

园冶访谈
INTERVIEW OF "YUAN YE AWARD" COMPETITION

随着2014年1月9日-10日园冶高峰论坛的成功举办，2013年"园冶杯"风景园林（毕业作品、论文）国际竞赛已完美地拉上了帷幕，来自国内外近百家风景园林相关院系的师生报名参加了本次竞赛。编者特别对部分优秀作品获奖者进行了采访，以期让大家更深入地了解优秀作品获奖者的设计经历和心路历程，为广大风景园林设计爱好者提供借鉴。

1. 您在整个课题参赛过程中是否有其他要分享的趣事？在您的设计之路上，有没有哪些人或事对您产生了很大的影响？

邵珊、李建磊：大山包国际重要湿地是黑颈鹤的主要栖息地之一，每年农历九月九到次年三月三，黑颈鹤会到此越冬，为赶在黑颈鹤离开前展开调研工作，我们大年初五便在两位老师的带领下来到了大山包。到达之后的第二天早上凌晨5点，我们便到大海子地下观鹤点守候黑颈鹤清晨起飞，日出伴随着栖息地鸟儿起飞的情景至今难忘。傍晚时分我们会在拥有绝对高度2600米绝壁峡谷的鸡公山守候日落，观日落坠入层层群山之中，然后在马帮的带领下骑着马儿回基地。在对当地社区居民的调研中，当地人很热情质朴，且为大山包作为黑颈鹤之乡而感到自豪，即使土地贫瘠，粮食产量很低，他们也愿意投食部分给黑颈鹤。调研时的引导员和我们讲，他去过很多地方，最终还是回到了家乡，觉得家乡最好，因为这里的食物绿色天然，更重要是高原精灵黑颈鹤令他想要永远留在这里守护他们。在回程的路上我们惊喜地发现了天然水晶，大家趴在跳墩湖湖岸石堆里刨水晶的那兴奋样儿，至今仍历历在目。这些难忘的经历都是我们在本次设计中收获的重要财富。

最初接触景观建筑设计时，就深爱赖特的设计，他提出的有机建筑概念提倡就地取材，强调建筑是有生命的、自然地从土地中生长出来，避免建筑与当地风貌格格不入的现象。能站在巨人的肩膀上了解"设计"是我们最大的快乐。感谢指导老师杨子江和李晖老师提供了这个有趣且有意义的课题，并指导我们完成设计，在毕业设计的整个过程中，两位老师都倾注了很多心血。获得荣誉之后反观当时才发觉，我们度过了本科设计学习过程中最充实、最有意义的最后半年。生活种种经历告诉我们设计是来源于生活的，我们对景观建筑设计的热爱和执着与各位老师、合作伙伴以及自身的经历密切相关，这一切是我们设计之路上最有力的支持。

2. 您认为您作品的闪光点在哪里？哪些方面还可以进行细化或者改进？

邵珊、李建磊：人类是自然的产物，最终也要回到自然中去，我们应遵循自然规律，加以合理利用。可持续问题经常被提到，但是由于人类的私欲和科技上的限制，很难将其付诸实践。我们认为设计师应该保持前卫理想化的头脑。就像本次设计的过程，是在一个理想化的前提下建立景观规划理论模型，最大优化湿地水域面积以及起到涵养水源作用的植被区域，接着不断地用现实约束条件，例如黑颈鹤的活动区域、习性、人类活动、干扰因素等等，不断修改模型，最终达到恢复湿地生态系统，建立人鹤共生理想环境的目的。我们坚信一个设计师不仅仅是一个理想家，更应该是一个社会学家，对人文的关怀。在本次设计中，起码我们在个人意识上保有了对社会和人类环境的责任感。

中国云南大山包国际重要湿地概念性景观规划是基于湿地生态系统恢复这个复杂而繁琐的研究过程，生态系统恢复是值得景观设计师深入研究的问题。由于毕业设计时间有限，一些专业问题还没有来得及深入研究，希望今后有机会和各个领域的研究人员多多交流。例如本次主题涵盖了关于黑颈鹤等野生动植物相互依存关系的生物学知识，关于风力水能等新能源利用的物理学知识，关于社区演化相关的社会学知识，以及生态学等各个领域的知识，这些都是需要细化和改进的地方。作为一名景观设计者，我们应吸收各个方面的信息知识作为设计过程中的学术理论基础。

陈荻、荣南：我们认为我们的作品值得细化和改进的地方还是不少的，主要是因为时间不足的问题，这也是我们这次设计的一个遗憾。比如拓展层中的景观构筑物形态设计还不够细致，拓展层设计的广义的方法论也显得有些欠缺；改造前后对比的效果图也只有4张，只是挑出了最主要的问题进行了说明与比较，一些有一定价值的细部没有完全展示出来等等。

如果一定要说闪光点的话，我们觉得还是我们设计过程中时间花的最足的部分，即老公园更新问题以及场地的调查研究，也正是因为花了时间，才看到了一些相对隐蔽的东西；通过这些调查研究，使得我们的研究有一定的科研性质，我们觉得这可能也是这次得奖的主要原因之一吧。

3. 您认为在设计过程中，首先要解决哪方面的问题？

陈荻、荣南：我们觉得设计过程的确是一个比较复杂的综合体，很多时候思维是感性的甚至跳跃的。在此次设计过程中，我们首先想要解决的是寻找出场地中存在的社会问题并分析与寻找对策，我们觉得这可能也是学科发展的重要趋势之一，而且社会问题的分析与解决是贯穿此设计始终的。

我们做的主题是老公园改造与更新，所以我们首先通过查找文献、实地调查等方法发现了目前老公园改造中存在的普遍问题，并提出了一种能够保留不同历史时段中文化与生态价值载体的更新方法，并制定出分层规划措施，并设定每层规划的方法，根据这些方法，把设计深化下去。

正是因为首先发现了场地中存在的社会问题并寻找了对策，整个设计在深化过程中也会显得目标明确，并相对易于开展。

4. 请问您设计方案的切入点在哪里？设计之初是如何构思立意并最终确定选题的？

张泽华：我设计方案的切入点是造成公园荒废现状的原因以及对周边人居环境所形成的影响。设计之初是通过对场地现场调查结合老师提供的地形，管线图等资料对照，对场地主要进行了趋于宏观与微观两方面的考虑：首先根据公园区位1000m服务半径内服务对象的整合定位公园性质，研究场地利用价值与景观的影响力，初步形成可持续性景观的中心立意；其次对场地高差、水系、高压走廊与荒废地貌等问题造成的原因进行微观层面了解分析，根据可持续性景观的主旨提出各项有效措施，参考文献资料最终形成景观处理手段。在此过程中提炼出打造城市中心绿洲、提倡低碳生活的立意构思。

5. 生态低碳型景观将成为我国未来城市建设的一大趋势，您认为可以采取哪些措施建设低碳、环保的景观？

张泽华：首先，生态低碳型景观随着城市建设发展有了越来越多有效可行的办法，它可以通过现代化科技改变景观材料，达到可持续利用、低排放的效果；也可以是各式环境技术的运用共同促进低碳，环保景观的建设，但是我认为作为未来的景观设计师，我们更有责任将低碳环保的景观概念带到我们的生活当中，而提倡一种积极向上健康活力的生活方式，应该将各式活动方式融入生态型景观当中，从而恢复公园真正的景观价值。

"园冶杯"获奖学生感言
PRIZE-WINNING STUDENTS' WORDS OF "YUAN YE AWARD" COMPETITION

"园冶杯"为不同地区开设相关专业的院校提供了一个展现自我的平台,对学生的激励作用不言而喻。将作品在网上公开以及在高校巡展,也让不同年级的学生了解到自己的优势与差距,最终共同提高研究和设计水平。

"Yuan Ye Award"Competition provides a platform for relevant professional institutions in different regions to show themselves, and it is an inspiring measure for students. Displaying online as well as exhibition tour in universities let students of different grades find out their advantages and disparities compared with others, and improve the level of research and design eventually.

规划设计其实是一个探索的过程,在理性与感性来回往复的过程中去寻求一个解答。竞赛其实也是一个摸索的过程,去理清题目的思路,去理清历来参与者的思路,去理清自己的思路,虽说参与竞赛最终目的应是明确而坚定的,但是全身心投入创作本身的那种全然不顾结果的设计冲动确是最难能可贵的回忆。作为年轻一辈的设计者,我们也将用我们的智慧为让世界变得更美好而不懈战斗。也祝"园冶杯"风景园林国际竞赛能承载更多更好的优秀设计作品奔向更明媚的未来。

——陈荻 荣南

首先感谢"园冶杯"风景园林国际竞赛组委会给予我们这次机会,以及对我们作品的肯定。这是一个很好的学习、交流和展现自己的平台。在我们的大学学习生活中需要这样的氛围,去执着地追求一个梦想,这样我们的大学生活才算充实。整个参赛过程是充实而美好的,这对即将踏入社会工作的我们是一个很好的锻炼,同时也为毕业设计画上了一个完美的句号。大赛举办的交流会汇聚了各大设计机构和景观大师,让我们学到了很多前沿的东西,也结交了很多业内的朋友。同时也希望大赛越办越好,能更深入地走进校园,不再局限于应届毕业生,让更多怀揣着梦想的学生走上这个学习交流平台。最后预祝2014"园冶杯"风景园林国际竞赛圆满成功!

——邵 珊 李建磊

首先非常感谢"园冶杯"风景园林国际竞赛组委会提供给我们一个展示自己学习成果的平台，其次非常感谢竞赛评委对我的设计作品的认可，也感谢给予我支持与鼓励的同学，还要感谢指导老师的指点和帮助。这次竞赛也让我收获不少，既展示了自己，又能认识到自己的不足之处，让我在今后的学习工作中有更明确的前行方向。最后，我祝愿"园冶杯"风景园林国际竞赛能够越办越好，提供更大的平台，发掘更多的优秀设计师，促进中国风景园林事业的蓬勃发展。

——张泽华

很荣幸在大学期间的最后一份作品可以获得这样的奖项，在整个规划设计过程中，我们结合实际情况对衰落的站区发挥了充分的想象，也在联系南京的本土特色和政府规划的同时不断地发现问题并解决问题。特别的感谢程云杉老师全程悉心指导，引导我们在更正确的方向上探索，更感谢"园冶杯"风景园林国际竞赛给我们提供这样一个展现自己才能的平台，这不仅是对我们作品的认可，更是对我们四年大学学习的认可。回顾付出与收获的整个过程，我们都更好地认识到了自己的优势与缺陷，相信从新的起点出发，我们会学到更多。谢谢！

——蒋沂玲 段倩 徐佳琪

在"园冶杯"风景园林国际竞赛中获得规划作品二等奖，我们真是倍感荣幸，参与比赛的过程，不仅是一种经验累积的学习，更是一种相互助长的机会，让我们体会到"景观"的深刻内涵。感谢评审老师们给予我们的肯定，这将会成为我们成长的一大动力，更要感谢此次比赛最大的功臣——戴宏一老师，感谢戴老师循循善诱地教导我们，不断与我们讨论和修正设计方案，才使得我们能获此殊荣。我们也在参赛过程中培养出了团队间的默契和合作情谊，感谢团队内所有人的努力，谢谢大家！

——陈冠云 梁韦茵

图中左起：梁韦茵、戴宏一老师、陈冠云

光阴似箭日月如梭，我在六年的时光之中完成了从本科到硕士的学习过程，并且顺利毕业。毕业设计是一个快乐与痛苦交织的过程，在这一过程之中我收获颇丰。各位老师和亲朋好友的支持和帮助给了我无限的动力。

"园冶杯"风景园林国际竞赛是一个很好的平台，这个平台让我们可以在一起交流各自的设计成果。在这一过程之中我们可以看到当代的风景园林事业有多么欣欣向荣。我很高兴参与其中，也很幸运得了奖，这也是评委老师对我的认可。我会继续加油，刻苦学习，为风景园林事业的发展贡献自己的一份力量。

最后，感谢陪我走过这六年的朋友们！也希望"园冶杯"风景园林国际竞赛越办越好！

——黄川壑

在大学生活的最后宝贵时光里，我们小组成员一起，将学生期间积累的知识和能力加以综合展示，将"园冶杯"风景园林国际竞赛作为重要目标，互相学习，合作并进，最终克服重重困难完成了设计作品，这个过程中我们培养了合作精神，碰撞出了设计灵感，收获了新的设计思路，感谢"园冶杯"风景园林国际竞赛给了我们一个尽情展现设计能力的舞台，也为我们在大学最后阶段的学习树立了关键目标。祝愿"园冶杯"风景园林国际竞赛越办越好，为更多优秀的大学生提供施展才华的舞台，收获更多优秀作品。

——高 枫 车建安 刘 熠 王 辉

感谢"园冶杯"风景园林国际竞赛丰富了我们大学最后一年的生活，依然怀念小组一起搜集资料、实地考察、研究方案、完成设计的美好时光，青春的汗水挥洒在收获的田野里。转眼间，年轻的我们正被镀上成熟的影子，人生之路在脚下延伸，希望中国举办更多这样的大赛，希望"园冶杯"风景园林国际竞赛越办越好！让更多园林人的青春被点亮！

——唐 菲 吴安琪 李旭彤

在学生阶段，我们能够将实际项目与竞赛结合的机会是非常难得的，"园冶杯"风景园林国际竞赛为我和队友提供了这样一个平台。让我们在满足施工、甲方等多方面的要求的同时，深刻地思考对环境对使用者有利的方面，并极力将其作为第一要务。"不积跬步，无以至千里"，改变是积小成多的，为了未来美好环境的改善，我们需要大刀阔斧地行动，更需要谨小慎微地思考和设计。我们切身体会到作为一名景观设计师，在拥有权利的同时，也背负着城市、自然、人类未来赋予我们的神圣使命。

——张 姝 杨文祺

参加了这次"园冶杯"风景园林国际竞赛之后，我们更加懂得如何去配合！感谢导师李敏教授的教导，感谢我的战友们！在人生的另一个起点，我们一起飞翔！

——陈磊晶 郑 懿 肖红涛

2013年，是一个改革之年、发展之年、创造之年、丰收之年、中国梦之年。"园冶杯"风景园林国际竞赛伴随我走过2013年，在这里收获了成绩，使我对设计的爱更加坚定不移。感谢我的指导老师张恒、刘仁芳、桑晓磊，有了他们的帮助我才拥有更强大的内心做出更完美设计。同时感谢"园冶杯"风景园林国际竞赛组委会的辛勤付出，让我在"园冶杯"的平台上更深刻地感受到了景观设计的魅力，望"园冶杯"风景园林国际竞赛伴随每一个有梦想的青年走过那最美好的一年！

——侯喆昊

参加这次"园冶杯"风景园林国际竞赛最大的收获是让我感受到了对景观设计师而言最为本真的梦想，那就是用自己的设计去给需要的人们带来便利的生活和更美好的生活环境，这就是我追求的全部。"园冶杯"风景园林国际竞赛给我的大学生活画上了完美的句点，这个奖项分量很重，让我更坚定地在自己的道路上前行。最后还要感谢"园冶杯"组委会，给了我们这样一个优秀的竞赛平台。更要感谢我的指导老师李辰琦教授对我的谆谆教诲和鼓励。

——贾晓丹

我非常高兴能参加这次"园冶杯"风景园林国际竞赛，一次经历，一次成长，在刚开始时并没有去想结果，只是抱着对专业知识的一种兴趣、一种求知心理，去参加比赛测试一下自己的设计能力及学习水平，并认真努力地做好每一件应该做的事情。能在这次竞赛中获奖倍感欣慰。这次竞赛的整个准备以及参赛过程对我来说无疑是一次美好的回忆。首先，我非常感谢孙振邦老师对我的热心帮助和悉心指导，感谢系里对这次竞赛的重视和大力支持。如果没有他们的努力和汗水，我们也不可能取得那么骄人的成绩。其次，我也非常高兴能加入我们的团队。团结就是力量，这次比赛就是对它最好的诠释。

——陆书雯

本次"园冶杯"风景园林国际竞赛最吸引我的地方就是它为景观设计界的新人们提供了一个展现自我的舞台，看到大家不同的设计理念和设计角度，都给我带来了很大启发，不仅仅在设计形式上有了改变，更重要的是提高了自己的意识。在设计中最关注的是对于解决问题的情感投入度，你所做的设计选择的是什么样的主题，要解决什么样的问题是你要特别关注的，在设计过程中如何处理好社会、美感和生态这三者的关系也是最关键的。要坚持把个人的情感注入到你的设计作品中去，才能得到一个自我的提升。

——高东东

在电脑作图已经完全普及的今天，主流做法是用软件渲染出华丽的平面图和效果图。我们希望能另辟蹊径，发挥同学们的特长，以手绘稿为主，去展现出我们的设计理念。由于我们并没有采用普遍的作图方法，因此很惊讶作品能在"园冶杯"中获奖。"园冶杯"风景园林国际竞赛为景观系的学生搭建了展示与交流的平台，使得我们能看到国内最优秀的景观毕业设计作品，也有机会得到不同专家的指导，丰富了我们的视野，也增加了我们的学识，希望"园冶杯"风景园林国际竞赛能够越办越好。

——刘悦 林黛

"园冶杯"风景园林国际竞赛是一个值得每一位风景园林专业的学子参加的专业性国际大赛。此次参赛让我感想颇丰，一次真正意义的拼搏与实践能让我们从中收获到平时在学校学习生活中难以得到的能力和经验。并且通过这个平台，让我有机会见识到不同院校对专业知识的见解和运用，以及专业评委对于作品的品评。希望接下来的参赛者们能够再接再厉！不断创新！祝愿"园冶杯"风景园林国际竞赛越办越好！

——孙林琳

时光荏苒，岁月如梭。转眼大学四年在一起奋斗的日子，随着这饱含我们激情和梦想的作品的完成而画上句号。回首过去，我们欢笑过，争论过，丧气过，坚持过，当我们终于把作品寄给"园冶杯"大赛组委会的那一刻，才有真实的满足感和成就感。感谢"园冶杯"给我们这次机会，让我们的努力得到肯定，同时也能从中发现自己与别人的差距。在此预祝2014"园冶杯"风景园林国际竞赛圆满成功，让我们看到更多更好的作品！

——余 梅 李今朝 李月文

通过参加本次"园冶杯"风景园林国际竞赛不仅学习了更多的专业知识，更重要的是开阔了视野。"园冶杯"也成为了各院校之间交流的重要平台，通过对其他获奖作品的学习，使我对该专业有了更多感悟，对于今后的学习起到了重要的作用。在此，希望"园冶杯"风景园林国际竞赛能够吸收越来越多的院校加入，创造出更多优秀的作品。

——李 杨

"园冶杯"风景园林国际竞赛以毕业设计作品为竞赛主体，对我们所有面临毕业的景观系学生来说都有着很大的吸引力，而期望在"园冶杯"中有所斩获也成为了我们做毕设时的一种动力。非常感谢"园冶杯"给我们学生提供了一个相互交流、切磋的平台，从别人的优秀作品中我们能看到自己身上的很多不足之处。景观路漫漫，愿我们能借助"园冶杯"的平台在未来的专业道路上越走越好。

——杨天人 常 凡 李 菁

能在本次"园冶杯"风景园林国际竞赛中获奖真的很让我们感到意外，本来一开始我们也就是抱着试试看的心态递交了作品，直到有一天有位同学突然跑来告诉我说在"园冶杯"官网上看见了我们的作品，我才想起这件事。很感谢"园冶杯"给我们颁发这个奖，让我们在以后的规划设计路上增添了许多自信，同时也让我们体会到了一个作品，从构思到设计再到参赛这一过程的艰辛。不过功夫不负有心人，我们的勤劳汗水最终还是换来了肯定和嘉奖，很欣慰也很庆幸。

——王艳秋 武 键

在进入风景园林专业学习的本科四年，学习途中也参加过几个大大小小的比赛，但是参加"园冶杯"风景园林国际竞赛一直是我在本科对专业追求的一个目标。"园冶杯"激励了很多同学，让一代又一代的园林人饱含着激情去创造。虽然本次作品只得到一个鼓励奖，但这足以说明参加"园冶杯"的作品都非常优秀，不同于国内其他很多的比赛。愿"园冶杯"风景园林国际竞赛能够使越来越多的园林人走向成功，成为中国风景园林发展史中的一座丰碑！

——周云婷

造园的乐趣在于通过感受而发现缺漏，然后通过不断地创想、尝试和实践，从而设计出以改善生活环境为目的的作品。感谢"园冶杯"风景园林国际竞赛这个专业平台，让我在设计作品得到认可之余，也为其他设计师们提供了表达个人理念的机会。与其说这是一场竞赛，不如说这是一次专业分享的难得的机会。因为我在这里发现，在设计中没有明确的界限用于划分不同的学术领域。我更相信，这是一个来自景观设计、城市规划、建筑和环艺等专业人士的分享会。

——曾舜怡

参与"园冶杯"风景园林国际竞赛的过程是让人难忘的，对于自己来说是一次宝贵的财富和经验。在设计项目的过程中，增强了自己的能力，会更加细心全面地考虑方案，非常认真地对待自己的设计内容。每一个方面和内容，都是自己亲自建立模型，绘出手绘方案，收获颇丰。感谢老师的悉心指导，感谢同学们的热心帮助，感谢这次竞赛让我得到了成长。希望"园冶杯"风景园林国际竞赛越办越好，让更多的学生可以通过这个平台得到锻炼和提升。

——马斯婷

"园冶杯"风景园林国际竞赛不仅促进了风景园林行业的和谐发展，提高了风景园林专业学生的创新意识及设计水平，活跃了学习气氛，展示了学子风采，同时也为毕业生们提供了一个全面展示才华和互相学习交流的平台。竞赛设置四类：风景园林设计作品类、风景园林规划作品类、园林规划设计论文类、园林植物研究论文类。每类设置两组：本科组和硕博组。我参加的是风景园林设计作品类，本科组。通过参加此次竞赛，我重新审视了自己的毕业设计作品——《沁园住宅区二期景观设计》，并重新整合，更加深入地了解了设计与生态的结合。感谢"园冶杯"给了我这次机会，也希望今后能有更多的同学参与到"园冶杯"风景园林国际竞赛中来。

——吴竹韵

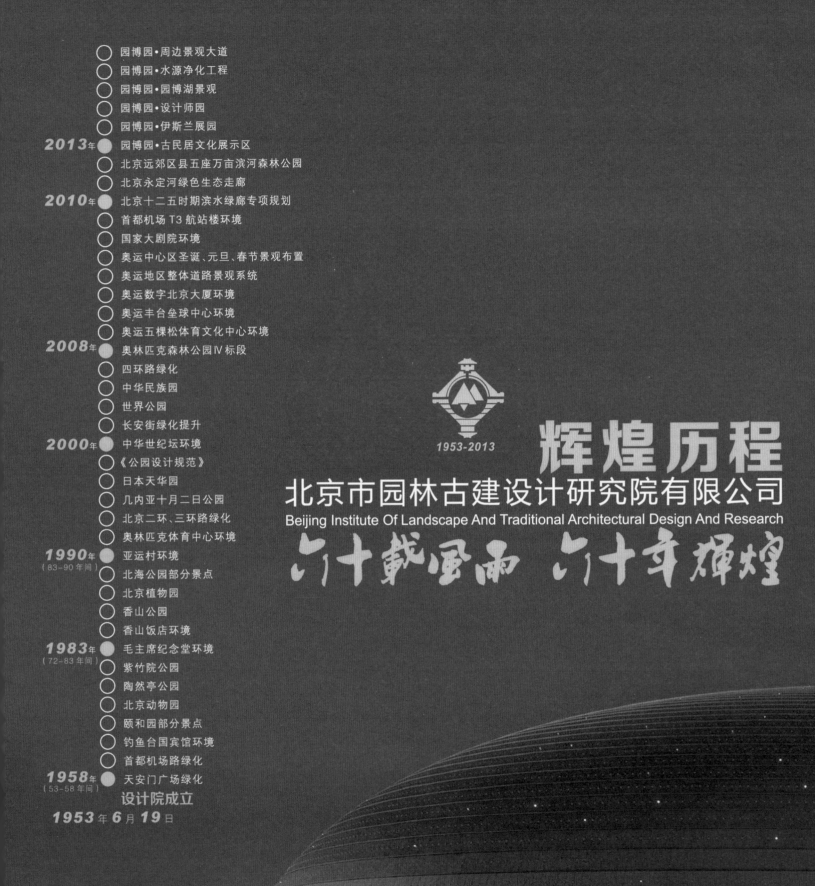

辉煌历程

北京市园林古建设计研究院有限公司
Beijing Institute Of Landscape And Traditional Architectural Design And Research

六十载风雨 六十年辉煌

1953-2013

- 园博园·周边景观大道
- 园博园·水源净化工程
- 园博园·园博湖景观
- 园博园·设计师园
- 园博园·伊斯兰展园
- **2013年** 园博园·古民居文化展示区
- 北京远郊区县五座万亩滨河森林公园
- 北京永定河绿色生态走廊
- **2010年** 北京十二五时期滨水绿廊专项规划
- 首都机场T3航站楼环境
- 国家大剧院环境
- 奥运中心区圣诞、元旦、春节景观布置
- 奥运地区整体道路景观系统
- 奥运数字北京大厦环境
- 奥运丰台垒球中心环境
- 奥运五棵松体育文化中心环境
- **2008年** 奥林匹克森林公园IV标段
- 四环路绿化
- 中华民族园
- 世界公园
- 长安街绿化提升
- **2000年** 中华世纪坛环境
- 《公园设计规范》
- 日本天华园
- 几内亚十月二日公园
- 北京二环、三环路绿化
- 奥林匹克体育中心环境
- **1990年** 亚运村环境
  （83-90年间）
- 北海公园部分景点
- 北京植物园
- 香山公园
- 香山饭店环境
- **1983年** 毛主席纪念堂环境
  （72-83年间）
- 紫竹院公园
- 陶然亭公园
- 北京动物园
- 颐和园部分景点
- 钓鱼台国宾馆环境
- 首都机场路绿化
- **1958年** 天安门广场绿化
  （53-58年间）
- 设计院成立
- **1953年6月19日**

A unified vision
To the visual
成就非凡 卓识远见

北京市园林古建设计研究院有限公司初创于1953年，是我国最早从事风景园林设计的单位之一，是第一批经建设部批准的"风景园林"甲级设计资质单位，同时拥有建筑工程乙级设计资质，具有规划咨询、园林设计、建筑设计等综合设计实力，可承揽相关领域的规划设计任务。

北京市园林古建设计研究院有限公司拥有一支集风景园林师、规划师、建筑师以及结构、给排水、电气、概预算等多专业工程师组成的160余人综合设计团队，其中拥有中高级技术职称人数近60%。

六十年来，**北京市园林古建设计研究院有限公司**凭借自身实力以及丰富的实践经验，始终处于行业中的领先地位，设计成果遍及国内外，如颐和园耕织图景区复建工程、北京奥林匹克森林公园Ⅳ标段施工图设计、国家大剧院景观工程、德国得月园、日本天华园等都得到了行业主管部门和社会人士的一致好评；多次在国家级、部级、北京市优秀设计和科技进步奖评选中获奖，累计达140余项。

北京市园林古建设计研究院有限公司是中国勘察设计协会常务理事单位、中国勘察设计协会园林与景观设计分会副会长兼秘书长单位、北京市勘察设计协会常务理事单位、中国风景园林学会理事单位、北京市园林学会理事单位，主持并参与了多项行业标准的制定，自行研发的园林规划设计软件在行业中广泛应用。

地址：北京市海淀区万寿寺路6号 电话：010-68423979 / 68423969 网址：www.ylsj.cn

人物专访：
EDSA Orient 总裁 李建伟
EXCLUSIVE INTERVIEW:
JIANWEI LI, THE PRESIDENT OF EDSA ORIENT

东方园林景观设计集团首席设计师	Chief designer of Orient Landscape
EDSA- 东方总裁	President of EDSA Orient
东方艾地景观设计公司总裁	President of Oriental Ideal Landscape Design Company
美国景观设计师协会会员	Member of American Association of Landscape Architects
美国注册景观规划设计师	American registered landscape designer
景观艺术专业硕士	Master of Landscape Art

李建伟先生是东方园林首席设计师、EDSA Orient 总裁。20多年的职业生涯中，李建伟不仅在城市规划和景观设计中建树卓著，也对高校园林教育有着亲身践行的经历，同时也担负着世界最大的园林企业、中国园林第一"股"——东方园林设计集团的高层管理。日前，本刊记者采访了这位"全才"型的专家，请他就园林专业学科建设、园林设计与施工，以及园林企业管理等方面，阐述自己独到而深刻的见解。

1. 在2013年"园冶杯"风景园林（毕业论文、作品）国际竞赛规划设计作品的评审会上，您对本科生的毕业作品有很多感慨，现场呼吁教育改革。请详细阐述一下您的观点。

李建伟：我觉得园林专业教育改革势在必行，尽管一些设计院对城市生态行业的发展中起着领头羊的作用，但是要改变行业的现状，最主要的还是靠教育。教育不改革，没有优秀的人才从学校里涌现出来，行业的发展就受到很大的制约。

中国目前的风景园林市场，在国际上是项目最多最有吸引力的，但遗憾的是设计师的总体水平比较低。随着现代城市的不断扩大，对生态的要求越来越高，景观行业囊括的领域越来越多，但学校的课程仍然没有改革，跟不上行业发展的需要。中国高校风景园林专业的知识结构亟需改革，应该根据市场的变化重新梳理相应的课程，向城市规划延伸，向历史人文延伸，向地理学延伸，向自然生态学延伸，向交通以及所有有关城市系统的领域延伸。

中国的景观设计教育太强调庭院艺术，忽视对技术、对科学以及对城市和自然生态体系的研究。其实，艺术与生态、科学是息息相关的。没有良好生态，就没有美的居住环境。生态、技术本身就是一种美，和艺术是不矛盾的。

中国的应试教育培养的学生接受知识能力强，而自主学习能力差。艺术类、设计类人才的培养要区别一般学生的培养，一定要给予他们自主创造能力发挥的空间，这样作为设计师才能放开束缚，有自己的想法，所以老师不仅仅是传播知识，而是培养学生主动学习的能力，指引方向，提供平台，培养学习能力才是教育的根本。

Jianwei Li is chief designer at Orient Landscape and president of EDSA Orient. Through his 20 years of practical experience, Jianwei Li has gained recognition from various circles in the planning and design industry. His achievements are evident in his commitment to the education of landscape architecture in major universities. Also, he plays a pivotal part in the senior management of Orient Landscape Group - the largest landscape architectural enterprise in Mainland China. a few days ago we conducted a face-to-face interview with this versatile expert. Li expressed some insightful views on the development of landscape education, landscape design and construction, and the management of a landscape design business.

1. During the 2013 "Yuan Ye Award" International Landscape Architecture Graduation Project/Thesis Competition, you passionately called for the reform of landscape design education when you saw the works (thesis and designs) of the young graduates. Is there anything in particular you want to address concerning this issue?

Jianwei Li: I think it is imperative for us to undergo a major reform of the educational system in the field of landscape design and planning. Although some design institutions have made important contributions in this area, education is the most crucial factor in the development of more talented people emerging from schools and therefore lessening the constrains on the development of this industry.

China has the largest and the most attractive market for landscape architecture but it is severely lacking in high-quality designers with comprehensive knowledge and skills. As contemporary cities continue to grow, people have higher

2. 您毕业于中南林业科技大学，之后在湖南师范大学艺术系主修美术，又在湖南大学建筑系主修建筑学，后来在美国获得景观设计硕士学位，您的求学经历、知识结构在国内设计师中并不多见，您当时就对自己的职业方向有了大致的规划？

李建伟：我小时候性格较内向，喜欢一个人呆在家里，偶尔看一本书引起了画画的兴趣，对艺术有了初步的领悟，酝酿了报考美术院校的想法。但高考恢复后，阴差阳错考到了中南林业科技大学林学专业，不过，机会总是给有准备的人，当中南林大准备重新组建园林专业时，一位北京林业大学园林设计专业毕业的老师在选修课上看中了我的美术功底，开始带我的毕业设计，并推荐我先后进修美术和建筑。毕业后我便参与了中南林大园林专业的组建，懵懵懂懂中走上了园林这条路。在中南林大八年的教书生涯中，我先后承担了张家界、韶山风景

expectations for the ecological condition of cities, which in turn, is leading the landscape design industry to become a more multi-disciplinary field of study. However the current course structure in our universities is still in some way old-fashioned, making it difficult for us to meet the requirements for the long-term development of this field. We need to make changes in the course structure that correspond to the changing market. I think more emphasis should be placed on urban planning, historical and cultural aspects, geography, urban ecological study and all the areas related to urban systems.

China has been focusing mainly on the tradition of garden art while neglecting a more scientific, technological approach to understanding cities and their ecosystems. We cannot see art as

图01 柳东大厦1
Fig01 Liudong building1

区等一系列湖南景观项目的资源发掘和规划。1992年，我已经通过了副教授的报批，但为了开阔视野，还是毅然选择到美国读硕士。这段留学经历颠覆了我原有的很多设计观念，使我真正理解了现代设计的内涵、语言和技巧。这段经历对我影响非常大，所以我经常建议年轻人能走出国门学习，从新的角度看待本土文化，学习新的知识。

3. 在2013年11月召开的北京园博会闭幕式暨颁奖大会上，东方园林因为建设皮特·沃克园获得施工大奖，这个小尺度只有3000平方米的园子，可能是东方园林做过的最小的项目，大师设计和东方园林施工，东方园林有哪些收获？极简主义的设计思想对当代青年设计师来说有哪些值得借鉴的地方？

李建伟：在2013年的第九届中国（北京）国际园林博览会上，

something contradictory to ecology and science. Art is closely related to science and ecology, as it is impossible for us to build up our living environments without an ecological perspective.

China's exam-based model strengthens students' abilities to acquire knowledge, but overlooks their capacity for self-directed learning. We should teach our students, especially those in the field of design, to be able to think independently and creatively rather than having them confined in a box that smothers creativity. The teacher's job is to establish a platform for students to learn how to learn and how to think.

2. You graduated from theCentral South University of Forestry and Technology, and majored in fine arts at Hunan Normal

图 02 柳东大厦 2
Fig02 Liudong building2

图 03 柳东大厦 3
Fig03 Liudong building3

美国景观设计大师皮特沃克携手东方园林奉献了"有限·无限"这一极具创意的园林佳作,东方园林也因该项目获得施工大奖。皮特·沃克花园是东方园林捐建的项目。皮特·沃克是世界上最有成就、最有影响力的景观设计师之一,承接这个项目对东方园林来说是一种荣耀。我们希望通过这个项目与大师建立良好的互动,将他的设计思想传播到中国设计师当中,并为公众奉献一处能成为亮点的花园。皮特·沃克的设计脱离传统范畴,应用大量虚幻景观,极具实验性。东方园林在这个项目上的投入、工程的细致程度和对大师设计理念的解读,也赢得了皮特沃克的赞誉,他评价说"这比我想象的还好"。

皮特·沃克是一位现代主义者,他经过长期的探索确立了自己的一套极简主义的设计手法。在园林创作中,多数人想要做到面面俱到、写实丰富。但把太多元素叠加入空间之中,元素的混杂反而会失去个性,流于平淡和模式化。艺术诉求的是"唯一性",一个作品不能追求包罗万象,简单纯净往往才能表达出特色和魅力。极简主义是我很欣赏的一种风格,在自己的设计实践中,从创意、铺装、栽植,都希望建立单纯、简洁、大气的风格。我主张设计师要追求个性,精研自己感兴趣的风格,反对中庸平衡之道。

4. 2013年下半年北京园博会、锦州世园会相继闭幕,这几年园博会、世园会、花博会非常热,很多地方政府都在积极争办。锦州世园会,东方园林从规划、设计到施工参与了很多,还有广西南宁园博会,也是东方园林涉足的一个类似的项目,您对这类项目如何评价?

李建伟:园林景观行业相对较小,公众的关注度不够,所以需要借助园博会、世博会这种形式来扩大行业影响,加强宣传,使更多民众意识到园林可以影响生活和城市,并且可以通过项目带动城市的区域发展。例如南宁园博会就有效带动了南宁市五象新区行政中心的发展,目前这一地区项目纷纷上马,假以时日就会建设成一个繁华市区。

目前此类展会还存在不少问题,除了显示各地传统外,展园设计缺乏创意或者急功近利,缺乏对园区地理环境的考察和合理规划,大拆大建,反而破化了当地的生态。明年将举行的青岛世园会就对崂山风景区的良好自然资源产生了一定程度的破坏。园博会的建设中,首先要遵循其宗旨,向市民宣传尊重自然、尊重生态的观念,改善人的生活环境,而不是大搞噱头,用奇奇怪怪的创意来夺人眼球,另外也展示园林设计的新理念、新技术。在南宁园博园的设计施工中,东方园林不但要考虑如何拉动周边地区的发展、如何建设更好的水与生态,还要重点关注园区的后续使用问题,在博览会的使命完成后,如何继续发挥其市民公园的功能。所以,好的设计不仅要具有前瞻性,权衡各方面的利益,还要重视其长效利用、后续发展。

5. 锦州世园会是一个以"海洋"为主题的滨海主题公园。这个项目带动一个新区的建设和环境的改善,这种运作模式是不是东方园林向城市景观综合服务商转型的一个范例?

李建伟:锦州世园会作为"中国第一个国家级海洋生态公园",由东方园林负责设计施工。这个项目是锦州市滨海新区最重要的公园,其目标就是带动一个区域的整体发展。作为景观综合服务商,东方园林已承建很多能够影响城市未来走向的"核心景观"——可称为城市"中央公园"的大尺度项目,在生态、空间布局、市民生活、区域开发等方面都带来很多重要功能和发展理念;同时,由单纯的景观建设走向为城市提供全方位服务,开发了生态、灯光、城市田园、城市森林、地产甚至婚庆等一系列的产业链。中国各地目前的设计水平参差不齐,往往会因配合上的不协调,破坏景观的整体效果。做城市景观综合服务商的目标是做出自己真正理想的作品。

6. 东方园林有将近600人的设计队伍,您领导着这样一个庞大的设计团队,同时您又是首席设计师,参与具体项目的设计工作,在设计和管理方面如何平衡?

University, and then studied architecture at Hunan University. You studied overseas and obtained your master degree in landscape architecture in the US. Your educational background significantly differs from other domestic designers. Could you please address your earlier career plans?

Jianwei Li: When I was a kid, I was a little bit withdrawn and liked to stay alone at home. It was a book that accidentally drew my attention to painting and initiated my interest and understanding of the arts and I started to have thoughts of going to art school. After the resumption of the university entrance exam, I entered the Central South University of Forestry and Technology based on a strange combination of circumstances. However, opportunities are always for those who are prepared. Some major changes happened to the course structure and I had the chance to undertake an elective subject and demonstrate my artistic skills. The tutor was really captivated by my works, and provided me with invaluable guidance on my final assignment and my future study in fine arts and architecture. After my undergraduate study, I participated in the reconstruction of the landscape architecture curriculum and commenced my career in this field. During my eight years of teaching at Central Southern Forestry University, I undertook a series of projects in Hunan including the scenic zones planning and resource management in Zhangjiajie, Shaoshan and so on. In 1992, I decided to go to the US for my master study and it was an enriching experience that broadened my horizons and completely changed my understanding of the meanings, language and techniques expressed in modern design. I am so grateful for the years I spent in US and I strongly encourage young people to study overseas to influence their views on local culture.

3. During the closing ceremony of the Beijing Garden Expo in November 2013, the construction award was conferred to Orient Landscape for the Peter Walker Garden. The garden is 3000 square metres in area, which I think is probably the smallest project for Orient Landscape ever. So, in what way has Orient Landscape benefitted from this project? And in what ways are young designers inspired by the ideas of minimalism and simplicity?

Jianwei Li: The design of Peter Walker's garden in the expo was based on his idea of 'finity and infinity' and Orient Landscape won the construction award for his masterpiece. The project was funded by Orient Landscape and it was an honour for us to work with one of the world's most successful and influential landscape architects on this project. I think working on such a project was a great opportunity for us to interact with a master of design and absorb his ideas from this distinctive garden. Peter Walker's design is opposed to conventional design approaches, in its creation of imagery through experimentation. In addition, we have to give credit to Orient Landscape for their attention to detail and their effort in interpreting Peter's design. We were happy to hear Peter say that the experience was better than he expected.

Peter Walker is a modernist who for decades has developed his own style of 'minimalism' and 'simplicity' through exploration. We have a number of designers, who make a great effort to achieve all-embracing designs in which numerous ideas and elements operate in tandem. I think a successful design is a simple

图 04 南宁五象湖 1
Fig04 Nanning Wuxiang Lake1

图 05 南宁五象湖 2
Fig05 Nanning Wuxiang Lake2

图06 南宁五象湖3
Fig06 Nanning Wuxiang Lake3

李建伟：设计师团队并没有复杂的人事关系和组织架构，我自身也是设计师中的一员，与团队从事相同工作，管理方面相对比较简单，要坚持亲力亲为做设计，有创意、有想法。只有自己身为一位优秀的设计师，才能服膺众人，更好地带领团队。我认为以员工能力和项目操作能够带动公司的发展，对管理层也要充分放权，给予他们足够的自主权和独立性。

7. 东方园林的设计集团未来几年有怎样的战略，对整个东方园林集团的设计施工一体化的发展模式有怎样的影响？

李建伟：东方园林设计集团一方面做好自身擅长的景观设计，另一面要立足于城市生态的协调发展。目前的趋势是，仅靠景观设计并不能解决城市生态和环境污染的问题。我们必须与水利、生态、环保、城市规划、建筑以及农业、旅游等多方面协同发展才能有效地解决城市的环境问题。对景观的理解也不再局限于植物、小品、水景等，而是人眼中所见都可归类广义上的景观，包括桥梁、道路、建筑、生态等等。国外完成一个项目，协调工作做得很好，而国内山头林立，彼此之间配合很困难，难以做出一个完整的、真正影响城市生态的好项目，这个问题过去十分令人头疼。如果以大型、综合性的设计院作为企业的发展方向，能避免项目建设上的支离破碎。东方园林目前已签约生态、水利、建筑、雕塑等各方面的专家，在同一家公司工作，更易配合协调，这样公司的实力越来越强，能更好地承担大型项目，服务于城市的发展。

8. 近几年您提出了"景观统筹"的概念，您操作过的项目中有类似的成功案例吗？景观统筹对于东方园林参与的大型市政项目有哪些积极的影响？对设计师的要求有哪些？

李建伟：在多年的项目实践后，我发现真正称得上成功的案例依然屈指可数，因为设计良好的景观往往被横插进去的道路、桥梁、周

and unique design that is distinct from others. We should have designers who practice to build up their own style rather than being confined within some old-fashioned doctrines.

4. The 2013 Garden Expo in Beijing and World Horticultural Exposition in Jinzhou were closed. Garden and horticultural expo events such as these which have been very popular in recent years and many local governments are actively bidding to hold such events. Orient Landscape was highly involved in the planning and construction of the Jinzhou Expo and also the Nanning Expo in Guangxi. What are your opinions on such projects?

Jianwei Li: The landscape industry is still a relatively constrained area that has not drawn enough attention from the community. This means we need more public events like garden expos to increase its influence on our society and make more people realize that landscape shapes their everyday life. The Nanning Garden Expo was a good example, as it spurred new development in the Wuxiang administrative centre, and I believe it will be turned into a bustling downtown one day.

However, problems still exist within such projects. There are concerns about lack of considerations of local community and ecological conditions due to our eagerness for short-term benefits. There are issues arising from lack of vision and appropriate planning as well. For example, the Qingdao Horticulture Expo development had some unexpected environmental impacts on the Laoshan natural reserve. I think these projects are platforms for representing our collective respect for the natural environments and the eco-system while improving the human

图 07 株洲神农城 1
Fig07 Zhuzhou Shennong City1

边建筑破坏了整体的美观，这成为困扰景观设计的一个难题。特别是在提出"城市生态"这个大的理念后，仅仅依靠绿化或园林设施并不能解决空气、水体等污染问题。不依靠景观统筹，就无法真正构建城市生态环境，给市民带来彻底的居住环境改善，从而离时代的要求越来越远。惟有景观才能协同多专业的发展，这是"景观统筹"这个理念提出的理论背景。现在国内的城市规划过于注重产业布局、交通桥梁、商业配套和绿地指标，而对资源禀赋、通风光照、人的活动方式等考虑不够充分。因此，城市各个部门缺乏沟通和统筹，往往是各干各的，结果景观也做了很多，但是却不够人性化，城市的整体景观没有体现出来．这就需要景观统筹规划，它就像是粘合剂，把各个部门整合到一起，绘制出一张生态宜居的蓝图，让景观融入到道路、建筑、桥梁等等所有的城市要素之中去。

为实践这个理念，设计师需要学习规划、生态、地理信息、城市历史人文等多方面的知识，介入到城市生态系统和整体规划中去，而不是仅仅为城市规划"填空"。我国的城市规划理论知识尚比较落后，例如湿地概念在专业人士中都缺少界定，缺少全国性的资源普查，各地也缺乏法规条例来保护湿地。行业中有很多这样的项目，为了造湿地，用上橡胶坝、防渗膜，结果就是假湿地假生态到处泛滥，导致大家被误导。

还有，地理信息是现代城市规划的一个重要依据，但是许多城市规划师和设计师都不了解地理信息的应用，也很少人能够操作相关软件进行分析、比较。我国目前能够使用的地理信息资源也比较缺乏，给城市规划统筹带来一定的难度。这些都是城市生态化实现的根本。

living environment. These events should also provide opportunities for us to raise innovative ideas related to landscape design. A successful expo project should produce long-term social and environmental benefits for the public and encourage the development of the surrounding areas.

5. The Jinzhou Expo is a seaside theme park based on the concept of 'ocean'. This development plays a key role in driving regional development and environmental improvement. Do you think this is a good example of how Orient Landscape can become more of a service provider that aims to introduce integrated urban landscapes?

Jianwei Li: The Jinzhou Expo project is thought to be the first national marine ecological park in China. Orient Landscape was fully involved in its planning and construction. The aim of this project is to establish a focal point that fosters development at a regional level. We work as a service provider to build more large-scale, central park landscapes in cities that serve urban ecology and public life and create opportunities for future regional development at a functional level. Also we are building up a new business model in which there are different sectors including ecology, lighting, garden cities, urban forests, property development and even wedding services and so on. Every area will be integrated with the landscape in our urban landscape production process.

6. Orient has a team of nearly 600 designers. You are the director responsible for managing a large group of people and also personally working on a large number of projects. How do you achieve a balance between these two different roles?

Jianwei Li: We work as a design team and are therefore involved with complex interpersonal connections and organizational structure. I consider myself as one of these designers and we are all working together to generate good ideas and produce quality designs. As a director of the team, you always need to make sure you have convincing skills and competency in order to better lead your team. I think it is essential to allow your team members space and freedom for individual thinking and expression.

7. What is the business strategy of Orient Landscape Group for the coming years? How will the strategy affect the current development model where design is integrated with construction?

Jianwei Li: Orient Landscape is focusing not only on planning and design, but also on achieving sustainable development for cities in the long term. We are faced with a lot of environmental issues and to have them resolved we must understand landscape as a multi-disciplinary field that is closely related to many areas such as city planning, architecture, ecology, water conservation, agriculture, tourism and so on. Our work-scope should be broadened and not limited to plants and water features that are merely pleasing to the eye. In developed countries, most of the projects are completed in a highly organised and systematic manner, whereas in China there are lots of projects that are fragmented and unsustainable due to the poor coordination from the beginning. In dealing with such issues, we have an advantage, as Orient is a large group encompassing experts working in diverse areas. Therefore I believe it is much easier for us to coordinate each aspect of the project, especially with large-scale projects.

8. In recent years, you have raised the concept of "landscape overall planning". Are there any successful cases in the projects you have operated? How does "landscape overall planning" actively affect the municipal projects which Orient Landscape has participated in? What does it require of landscape architects?

Jianwei Li: After so many years working on projects, I find that the cases considered to be successful are still few and far between. This is because well designed landscapes are usually disturbed by roads, bridges and surrounding buildings, and then lose their overall aesthetic effect, this is a serious problem in landscape architecture. This is particularly evident when considering the idea of "city ecology". We can not solve problems such as air and water pollution through greening or landscape construction alone. Without landscape overall planning, we can not truly

图 08 株洲神农城 2
Fig08 Zhuzhou Shennong City 2

Fig09 Zhuzhou Shennong City3

construct urban ecological environments that fully improve urban living conditions, but will rather move further and further away from the relevant environmental issues of the times. It is only through landscape that we can coordinate the development of multiple professions. This is the theoretical background for the concept of "landscape overall planning". Nowadays in our country, urban planning pays too much attention to industrial layout, roads and bridges as well as commercial configurations and green space norms, but does not pay enough consideration to resource endowment, ventilation, illumination, human behavior and so on. Therefore, all city departments lack communication and overall planning with one another. As a result, we have built so many landscapes that are short of humanity, and the city's overall landscape potential does not emerge. Therefore, landscape overall planning is necessary. It is like a glue, that brings together all departments to draw a blueprint for ecological livability, and merges landscape into all urban elements, such as roads, buildings, bridges, etc.

In order to practice this idea, architects need to become knowledgable on planning, ecology, geographic information systems, and urban history and culture, and integrate themselves into urban ecosystems and overall planning rather than just fill holes the in urban planning process. In our country, the theoretical knowledge on urban planning is still poor. For example, the concept of wetland is not defined by our professionals, we lack general surveys on national wetland resources, and no district has produced comprehensive laws and regulations on wetland protection. There are many projects that utilize rubber dams and impermeable membranes to construction wetland. As a result, fake wetlands and false ecologicalization overflows everywhere, misleading people.

Moreover, geographic information is an important basis for urban planning nowadays. However, most urban planners and architects don't know how to make use of geographic information systems, and few are able to operate related software for comparison and analysis . At present, our country lacks available geographic information resources, and therefore has difficulty with urban planning. This information forms the basis for realizing urban ecologicalization.

人物专访 EXCLUSIVE INTERVIEW

作者简介：

1995年毕业于美国明尼苏达大学，获景观艺术硕士学位，次年加入国际著名规划设计公司——美国EDSA，在美洲、欧洲、亚洲、中东等多个国家有着丰富的跨国设计及从事教育培训的经历，后凭借杰出的实力被擢升为EDSA合伙人。2011年李建伟凭借在景观设计界的超高影响力以及在区域景观规划、城市景观系统、高档主题酒店、旅游度假项目等领域的卓著建树，受邀成为中国园林第一股——北京东方园林股份有限公司的首席设计师。该公司成立于1992年，是集设计、施工、苗木、运营、养护全产业链发展的城市景观系统运营商。其中，景观设计集团拥有EDSA-东方、东方利禾、东方艾地、东联设计四个著名设计品牌。

Biography:

Jianwei Li graduated from the University of Minnesota in the United States with master's degree in landscape architecture in 1995. The following year, he joined EDSA, a renowned planning and design company. After completing a range of high profile design projects, he was promoted to a partner. Due to his contributions to the landscape design profession, particularly in the fields of regional landscape planning, urban landscape system, and resort and tourism landscapes, Jianwei Li was invited to be the chief designer of the Beijing Oriental Garden in 2011, by the development company Share Ltd. The company, founded in 1992, is engaged in the full lifecycle of urban landscape development, from the design process through to project management and construction. The Share Ltd. Landscape design group operates under the design brands EDSA Orient, Orient LIHO, Oriental Ideal, and OUDG.

查尔斯·沙（校订）
English reviewed by Charles Sands

竞赛作品
CONTEST ENTRIES

一等奖 / THE FIRST PRIZE

中国·云南·大山包
国际重要湿地概念性景观规划设计
DASHANBAO INTERNATIONAL WETLAND CONCEPTUAL LANDSCAPE PLANNING DESIGN IN YUNNAN, CHINA

中文标题：中国·云南·大山包国际重要湿地概念性景观规划设计
组　别：本科
作　者：邵 珊　李建磊　恭珉皓　孙 计　张冬妮
指导教师：杨子江　李 晖
学　校：云南大学
学科专业：景观建筑设计
研究方向：生态景观
分　类：2013"园冶杯"风景园林（毕业作品、论文）国际竞赛规划类作品一等奖

Title: Dashanbao International Wetland Conceptual Landscape Planning Design in Yunnan, China
Degree: Bachelor
Author: Shan Shao, Jianlei Li, Minhao Gong, Ji Sun, Dongni Zhang
Instructor: Zijiang Yang, Hui Li
University: Yunnan University
Specialized Subject: Landscape Architecture Design
Topic: Ecological Landscape
Category: The First Prize in Planning Group of the 2013 "Yuan Ye Award" International Landscape Architecture Graduation Project/Thesis Competition

作品简介：

　　本设计通过概念性景观规划对大山包湿地景观生态系统进行恢复，从而构建人、鹤、自然共生的生态体系（图01）。

　　基于现状调查研究分析（图02-04），利用GIS多因子叠加评价体系，分析和评估湿地景观生态系统现状，通过数据可视化的方法与手段，展现科学完整的分析过程，试图建立一套完善的湿地景观恢复规划理论模型。

　　首先设定在理想状态下，基于现状水系及湿地形成条件，进行湿地面积最大化规划（图05）；由于植被是湿地涵养水源的重要因子，通过分析当地林种生长条件（坡度、坡向、海拔等），水源点林地涵养保护，结合现状林地构建最优林地分布图，从而产生一个新的湿地数据层（图06）；其次以黑颈鹤活动范围因子（亲水范围、水深、人类干扰等）对其他因子进行控制（图07），然后结合居民点因子和农田因子，不断调整（图08），得出各类专项规划图（图09-12），最终得出规划总图。简而言之，即在湿地恢复规划中先做"加法"后做"减法"的逆向思维。这一方法适用于因子关系复杂的景观生态体系规划中，逐层研究分析因子间的影响关系，构建规划理论模型。

Introduction:

　　The project aims to restore the ecological system of wetland landscape via the conceptual landscape planning, and thus build a symbiotic system with people, wild animals and organic nature (Fig.01).

　　Based on the investigation and analysis (Fig.02-04), we analyze and evaluate the present situation of Dashanbao wetland landscape ecosystem by using GIS superimposition of multiple factor evaluation, which shows the scientific analysis process through methods of data visualization. Finally, we attempt to establish a comprehensive set of wetland landscape restoration planning theory models.

　　Firstly, this plan maximizes the wetland area based on the present situation of the water system and the wetland formation conditions (Fig.05). Secondly, vegetation is an important factor of water conservation in a wetland, so we plan the optimal forest distribution through analyzing the local plants' growing conditions (slope, aspect, elevation, etc.) as well as the protection of headwaters, and then create a new layer of wetland based on this information (Fig.06). Thirdly, the other aspects are controlled by the Grus nigricollis' range of activities (hydrophilic scope, depth of water, human disturbance, etc.) (Fig.07). Combined with settlements and farmland factors and with constant adjustment(Fig.08), we can draw an assortment of special plans (Fig.09-12). And ultimately a general layout plan is completed. In short, it can be understood as reverse thinking in the wetland restoration planning process, in which the "subtraction" is done after the "addition". By analyzing the influence between two factors step by step before constructing the planning theory model, the system of landscape ecological planning is approached with complex relationships among factors.

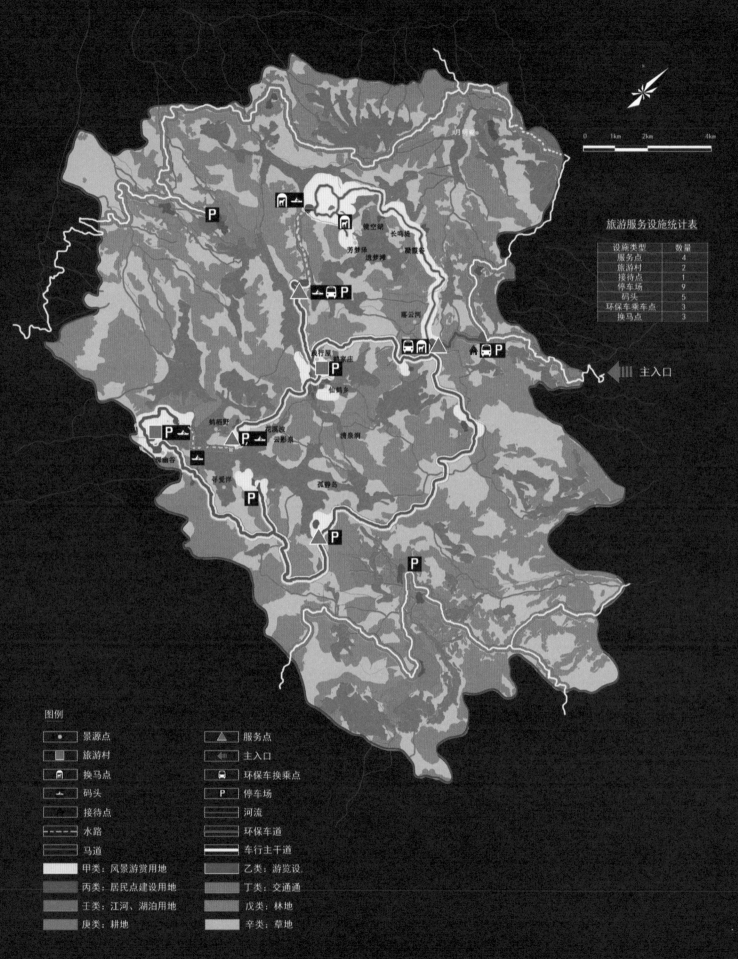

图01 规划体系结构及规划总平面图
Fig01 Master plan

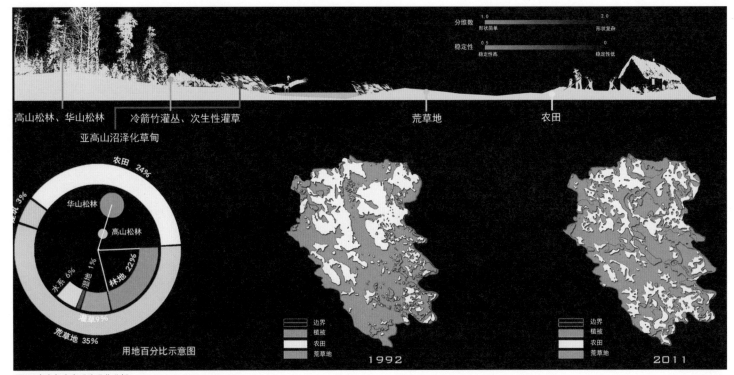

图 02 大山包生态系统退化分析
Fig02 Analysis of dashanbao ecological system degeneration

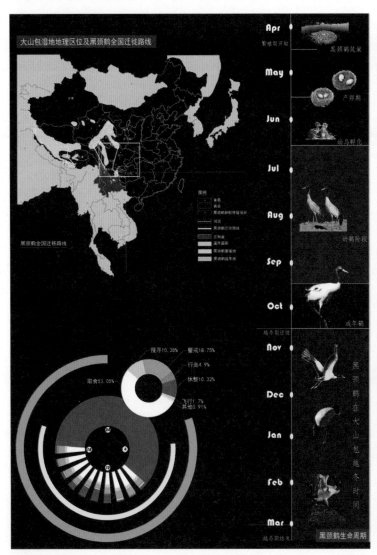

图 03 黑颈鹤迁徙及生命周期分析
Fig03 Analysis of black-necked cranes' migration and life cycle

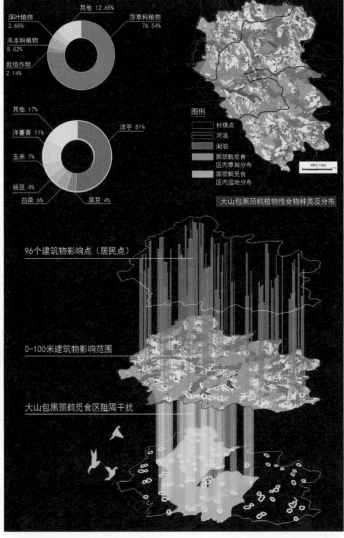

图 04 黑颈鹤食物及人类干扰空间分析
Fig04 Analysis of black-necked cranes' food and human interference space

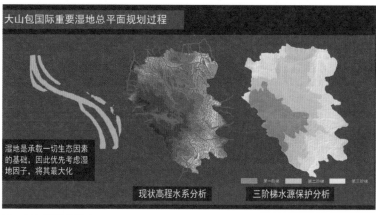

图 05 最大化湿地规划图
Fig05 Planning graph of maximize wetland

图 06 最大化林地规划图
Fig06 Planning graph of maximize forest land

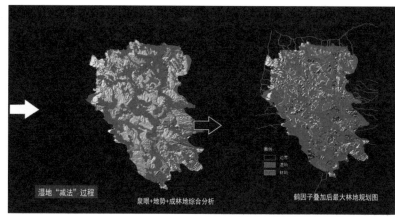

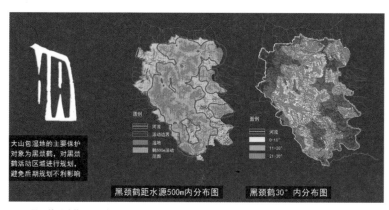

图 07 黑颈鹤活动范围因子控制规划图
Fig07 Planning graph of black-necked cranes' activity area factor control

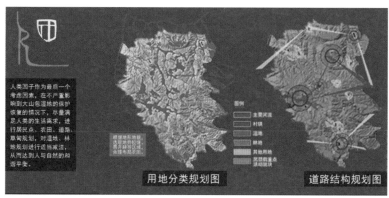

图 08 人类因子叠加规划图
Fig08 Planning graph of human factor superposition

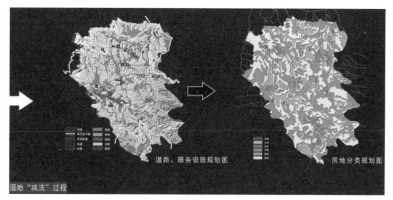

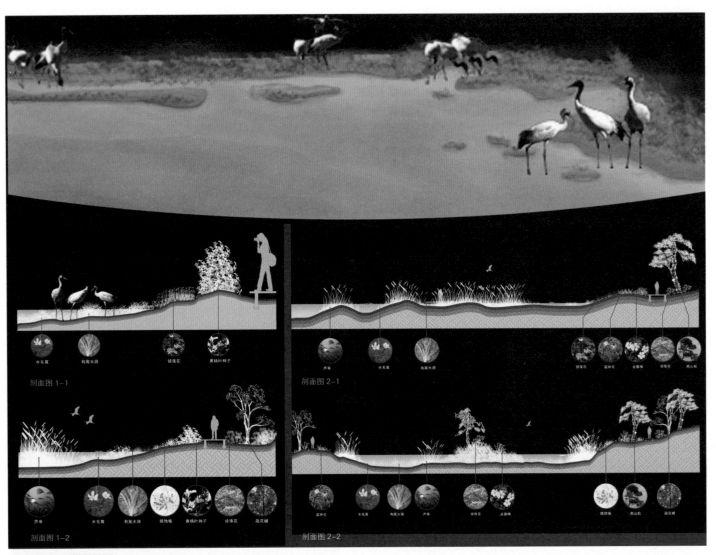

图 09 湿地生态驳岸景观设计
Fig09 Design of wetland ecological revetment

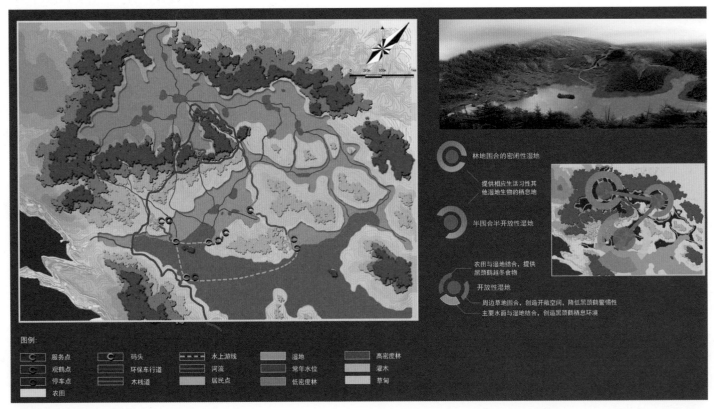

图 10 重要节点概念性景观设计
Fig10 Conceptual landscape design of important nodes

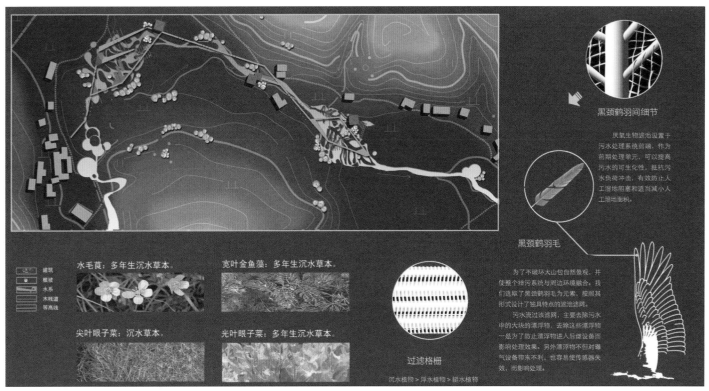

图11 概念性景观污水处理
Fig11 Conceptual sewage treatment

图12 观鸟景观设施概念设计
Fig12 Conceptual design of bird watching landscape facilities

查尔斯·沙(校订)
English reviewed by Charles Sands

老公园更新演绎
——以黄兴公园改造为例
THE PROGRESS OF UPDATING AN OLD PARK
——TRANSFORMATION OF HUANGXING PARK, SHANGHAI

中文标题：老公园更新演绎——以黄兴公园改造为例
组　　别：本科
作　　者：荣　南　陈　荻
指导教师：邱　冰
学　　校：南京林业大学
学科专业：风景园林学
研究方向：景观设计
分　　类：2013 "园冶杯" 风景园林（毕业作品、论文）国际竞赛设计类作品一等奖

Title: The Progress of Updating an Old Park——Transformation of Huangxing Park, Shanghai
Degree: Bachelor
Author: Nan Rong, Di Chen
Instructor: Bing Qiu
University: Nanjing Forestry University
Specialized Subject: Landscape Architecture
Topic: Landscape Design
Category: The First Prize in Design Group of the 2013 "Yuan Ye Award" International Landscape Architecture Graduation Project/Thesis Competition

作品简介：

本设计通过对黄兴公园的区位与现状分析，指出了"有机更新"与"普通更新"效果的区别，提出了一种通过设置基底层、保护层和拓展层对黄兴公园各历史分层进行统筹规划的策略，形成了一种以分层思想为指导的公园有机更新模式：将黄兴公园各历史分层中具有历史文化、生态及使用功能价值的部分视为一个连续开放的整体划入基底层、保护层，包含节点改造与保留（图01-04）；以拓展层来提升黄兴公园的整体功能以适应新的发展，主要手段包括渗透外圈（图05-06）、环形跑道及其立剖面、人行天桥的设计（图07）；从而使得黄兴公园整体功能提升，并有效进行有机更新（图08-11）。

Introduction:

The program points out the difference between "organic renewal" and "standard renewal"(Fig.03) through an analysis of Huangxing Park. The overall-planning strategy is based around identifying a symbolic 'basal layer', 'protective layer' and 'expanding layer', and designing them based on corresponding historical stratifications. This strategy forms a model of organic renewal for Huangxing Park, based on hierarchical thinking. The elements of history, culture, ecology and function are placed into the basal layer while the protective layer, contains a reserve and a transformation of nodes, which act as a continuous open whole(Fig.01-04). The expanding layer is set to enhance the overall function of the park and adapt to new developments, including penetration from the edges(Fig.05-06). This layer contains a running track and a pedestrian bridge (Fig.07). All of these strategies will promote the effective overall- renewal of Huangxing Park(Fig.08-11).

图01 节点保留
Fig01 Reserve of nodes

图02 节点改造
Fig02 Transformation of nodes

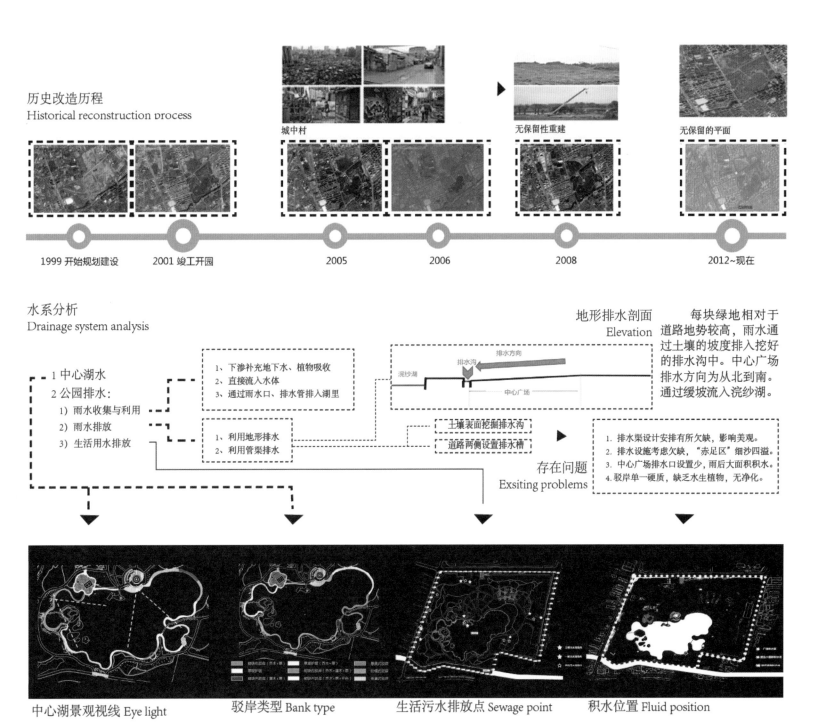

图03 现状分析
Fig03 The analysis of current situation

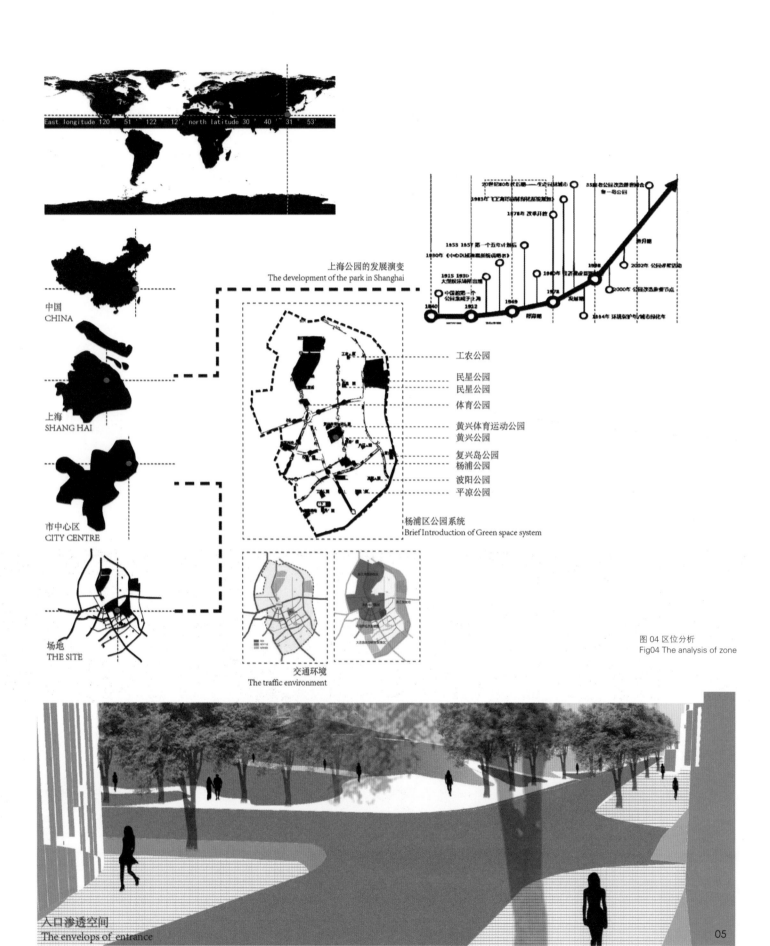

图 04 区位分析
Fig04 The analysis of zone

竞赛作品 CONTEST ENTRIES

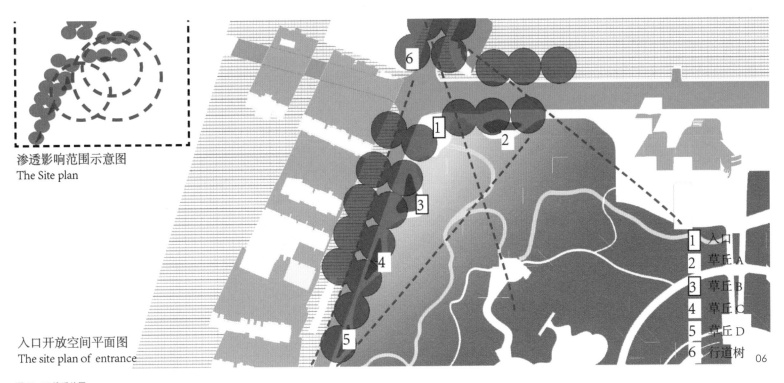

渗透影响范围示意图
The Site plan

入口开放空间平面图
The site plan of entrance

1 入口
2 草丘A
3 草丘B
4 草丘C
5 草丘D
6 行道树

图 05-06 渗透外圈
Fig05-06 Outer penetration

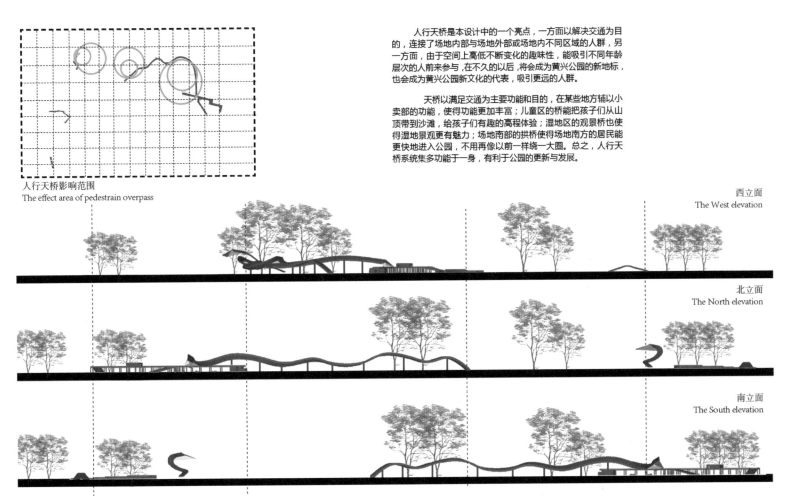

人行天桥影响范围
The effect area of pedestrain overpass

人行天桥是本设计中的一个亮点，一方面以解决交通为目的，连接了场地内部与场地外部或场地内不同区域的人群，另一方面，由于空间上高低不断变化的趣味性，能吸引不同年龄层次的人前来参与，在不久的以后，将会成为黄兴公园的新地标，也会成为黄兴公园新文化的代表，吸引更远的人群。

天桥以满足交通为主要功能和目的，在某些地方辅以小卖部的功能，使得功能更加丰富；儿童区的桥能把孩子们从山顶带到沙滩，给孩子们有趣的高程体验；湿地区的观景桥也使得湿地景观更有魅力；场地南部的拱桥使得场地南方的居民能更快地进入公园，不用再像以前一样绕一大圈。总之，人行天桥系统集多功能于一身，有利于公园的更新与发展。

西立面
The West elevation

北立面
The North elevation

南立面
The South elevation

图 07 人行天桥
Fig07 Pedestrian bridge

图 08 分层策略
Fig08 Strategy based on hierarchical thinking

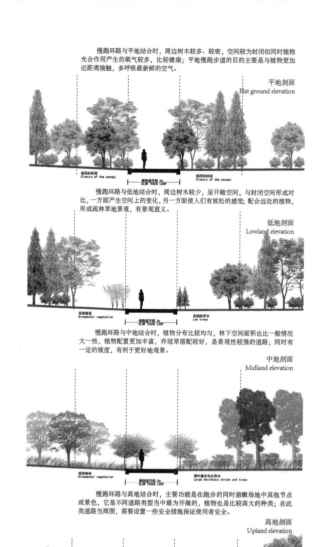

图 09 环形跑道及其立剖面
Fig09 Running cycle and it's sections

图 10 鸟瞰图
Fig10 Bird's eye view

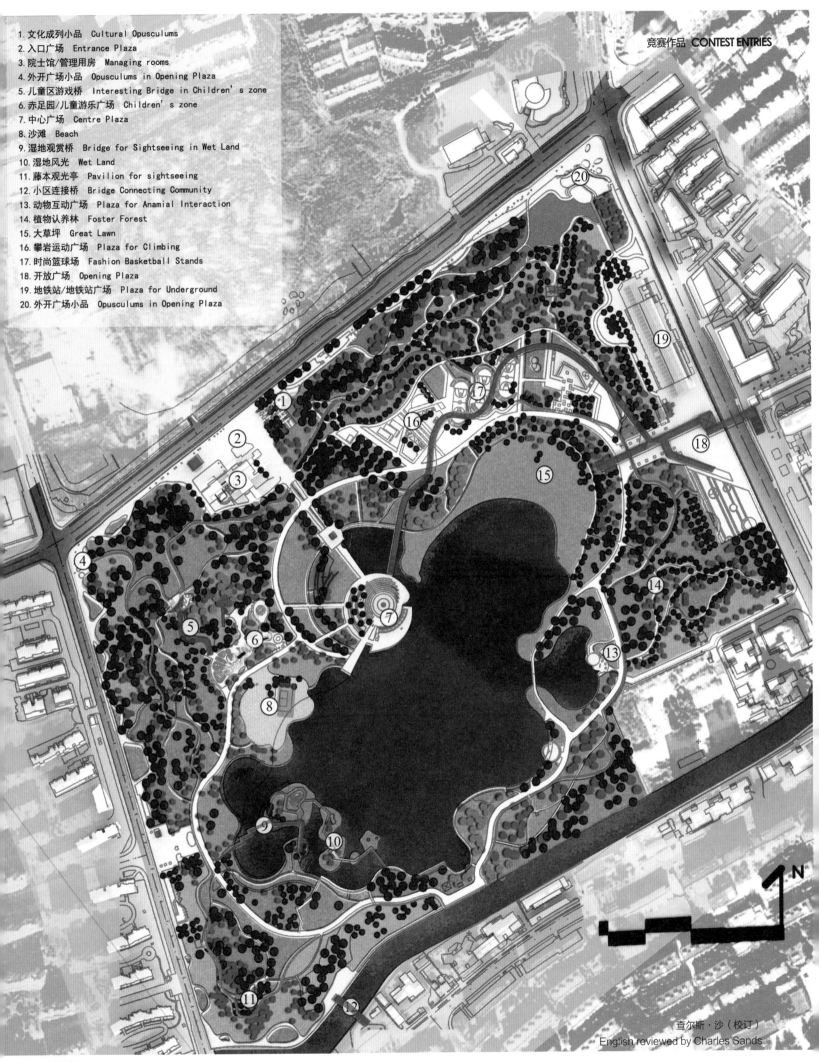

Fig11 Master plan

1. 文化成列小品 Cultural Opusculums
2. 入口广场 Entrance Plaza
3. 院士馆/管理用房 Managing rooms
4. 外开广场小品 Opusculums in Opening Plaza
5. 儿童区游戏桥 Interesting Bridge in Children's zone
6. 赤足园/儿童游乐广场 Children's zone
7. 中心广场 Centre Plaza
8. 沙滩 Beach
9. 湿地观赏桥 Bridge for Sightseeing in Wet Land
10. 湿地风光 Wet Land
11. 藤本观光亭 Pavilion for sightseeing
12. 小区连接桥 Bridge Connecting Community
13. 动物互动广场 Plaza for Anamial Interaction
14. 植物认养林 Foster Forest
15. 大草坪 Great Lawn
16. 攀岩运动广场 Plaza for Climbing
17. 时尚篮球场 Fashion Basketball Stands
18. 开放广场 Opening Plaza
19. 地铁站/地铁站广场 Plaza for Underground
20. 外开广场小品 Opusculums in Opening Plaza

查尔斯·沙（校订）
English reviewed by Charles Sands

城市绿洲，低碳生活
——重庆南岸区兰花湖公园景观规划设计
URBAN OASIS, LOW-CARBON LIFE
——LANDSCAPE DESIGN OF ORCHID LAKE PARK IN NANAN DISTRICT, CHONGQING

中文标题：城市绿洲，低碳生活——重庆南岸区兰花湖公园景观规划设计
组　　别：本科
作　　者：张泽华
指导教师：罗爱军
学　　校：西南大学
学科专业：风景园林
研究方向：景观规划设计
分　　类：2013"园冶杯"风景园林（毕业作品、论文）国际竞赛设计类作品一等奖

Title: Urban Oasis, Low-Carbon Life——Landscape Design of Orchid Lake Park in Nanan District, Chongqing
Degree: Bachelor
Author: Zehua Zhang
Instructor: Aijun Luo
University: Southwest University
Specialized Subject: Landscape Architecture
Topic: Landscape Planning and Design
Category: The First Prize in Design Group of the 2013 "Yuan Ye Award" International Landscape Architecture Graduation Project/Thesis Competition

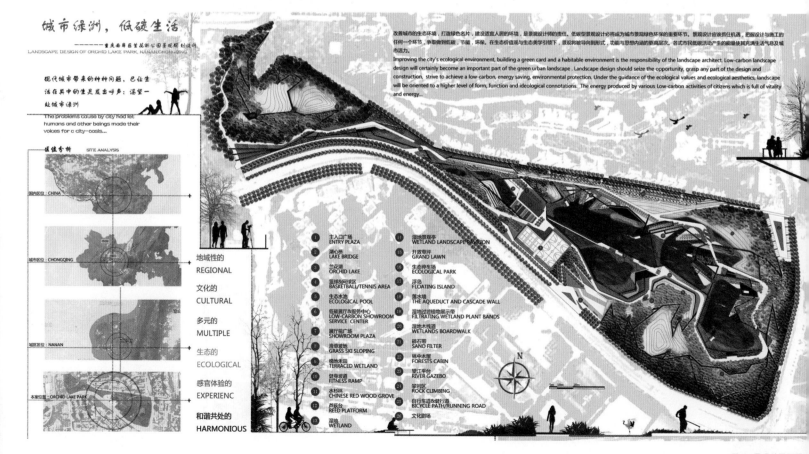

图01 景观总平面图
Fig01 Master plan

作品简介：

本方案通过提出以过滤汇聚中水为主的方式收集与储备湖水，从而建立兰花湖公园内湖湿地，同时保留当地特色山地农业印象，规划改建现有场地形成现代线性环湖系统，并融入低碳健康的生活方式方法以解决兰花湖汇入量不足造成常年干涸，场地存在较大高差，以及现状荒废地面貌等一系列问题（图01-04），并在此基础上对公园进行景观规划设计，进而形成集休闲，娱乐，运动，生态，生产及教育于一体，现代设计与传统生活方式碰撞与融合的城市"绿洲"（图05-08）。

Introduction:

The design puts forward a method of filtering and gathering recycled water in order to establish a wetland in Orchid Lake Park and preserve the local characteristics of the original mountain agriculture. The plan involves rebuilding the existing site to form a modern linear system around the lake. Low carbon elements and processes are integrated throughout to create a healthy environment for people and mitigate the problems of Orchid Lake. These include insufficient infiltration, major elevation differences and contaminated land in the park (Fig.01-04). Based on the above issues, the landscape of Orchid Lake Park is planned and designed to combine leisure, entertainment, sports, ecology, and education into an organic whole to form an "Urban Oasis". It is an amalgamation of modern design with a traditional way of life (Fig.05-08).

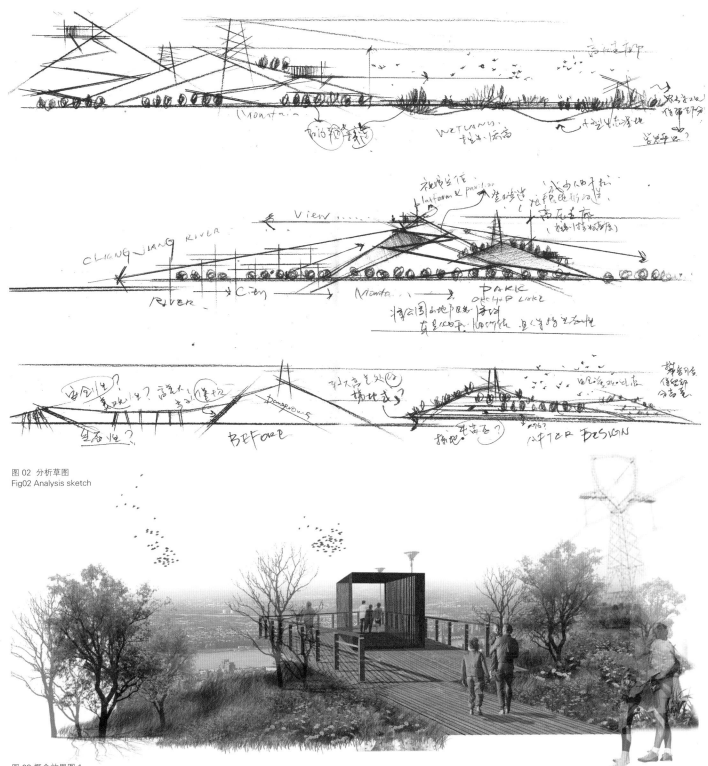

图02 分析草图
Fig02 Analysis sketch

图03 概念效果图1
Fig03 Concept perspective1

❶ 场地多处存在较大高差。规划区内现状高程最低点为243.09米,最高点为296.03米,竖向高差约50米左右。使用范围较小景观效果较差,并给周边居民进入公园造成不便。

兰花湖现状汇入量不足,常年干涸,面积约21000平方米,且园内缺少有效的管理,用地多为居民"菜园",各式污染严重,使生态系统遭到破坏。

公园周边开发建设后,形成城市荒弃地面貌,公园的荒废对入园者造成一定的安全隐患,丧失了城市中心绿地的作用功效

❷ 针对兰花湖现状问题提出初步改造建议,有序合理地整理解决汇水量少,高差大,缺管理,污染重等多个主要问题。同时重新梳理公园种植,使原本的"贫瘠之地"成为未来的"城市绿洲"注入低碳健康的生活方式理念,使公园逐步成为新型城市中心绿地

❸ 在生态价值观与生态美学的引领下,景观将被导向到形式,功能与思想内涵的更高层次,且各式居民低碳活动产生的能量使其充满生活气息及城市活力

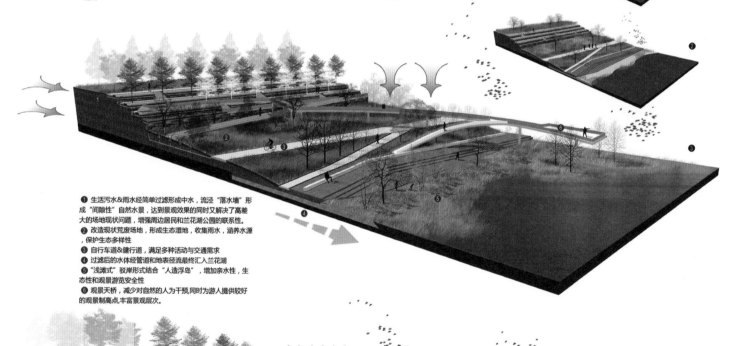

❶ 生活污水&雨水经简单过滤形成中水,流径"落水墙"形成"间隙性"自然水景,达到景观效果的同时又解决了高差大的场地现状问题,增强周边居民和兰花湖公园的联系性。
❷ 改造现状荒废地,形成生态湿地,收集雨水,涵养水源,保护生态多样性
❸ 自行车道&健行道,满足多种活动与交通需求
❹ 过滤后的水体经管道和地表径流最终汇入兰花湖
❺ "浅滩式"驳岸形式结合"人造浮岛",增加亲水性,生态性和观景游览安全性
❻ 观景天桥,减少对自然的人为干预同时为游人提供较好的观景制高点,丰富景观层次。

图 04 湿地区分析
图 04 Wetland analysis

图05 概念效果2
Fig05 Concept perspective 2

湿地休闲区内5处景观亭由不同的低碳材质（木，竹，砖，石，金属）为材料构成。
水平视线方向：以人水平视锥角度平均60度为参照，布置五处景观亭位置，使五处景观亭视线尽观整个湖区，达到较好的景观效果。
垂直视线方向：考虑个别景观亭内座位高度控制，以垂直视锥范围30度为参考，并结合景观亭高度2.0-2.4M的设计，尽可能将坐下休息的人的视线避开山地高压走廊。

Five pavilion built with five kinds of low carbon materials (Bamboo,Wood,Brick,Stone and Metal)
Horizontal line of sight direction:Cone Angle on average 60 as reference, arrange the pavilion to make sight view across the lake.
Vertical direction:Considering individual seat height of pavilion, vertical cone angle 30 as reference,with the design height of 2.0-2.4M,to make the view avoid the high voltage corridor.

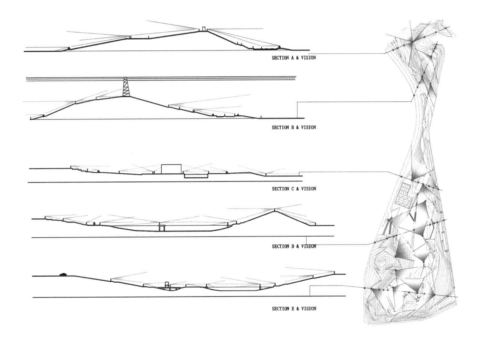

图 06 视线分析
图 06 Vision analysis

图 07 分区鸟瞰1
Fig07 Partition bird's eye view 1

图 08 分区鸟瞰 2
Fig08 Partition bird's eye view 2

在湖边坐下，静静地看眼前的一切，不受打扰地享受红尘俗世之外的淡定……

查尔斯·沙（校订）
English reviewed by Charles Sands

昆山夏驾河"水之韵"城市文化休闲公园（廊架小品）
——2013全国优秀工程勘察设计行业园林景观三等奖、2013年度上海市优秀工程设计一等奖

上海亦境建筑景观有限公司以规划、建筑、景观专业的设计实力为依托，构建了从项目前期策划、整体规划、建筑设计、景观设计到景观工程的一体化服务体系，为客户提供项目前期至后期的全方位解决方案。

亦境公司汇聚百余名设计精英、驰骋行业二十年，拥有建筑行业乙级、风景园林专项甲级、城市园林绿化二级等行业资质，作品多次获得国家建设主管部门及上海市优秀工程设计奖项。

大型城市滨水区
昆山夏驾河"水之韵"城市文化休闲公园规划设计
昆山夏驾河滨水商业娱乐区规划、建筑、景观一体化设计
泰州凤城河-西南城河景观、建筑一体化设计
镇江古运河风光带总体规划及分段设计
镇江国宾馆建筑、景观、夜景灯光一体化设计（金山湖）
镇江滨江风光带规划、建筑、景观、灯光一体化设计
海南博鳌香槟郡规划与建筑设计（滨海综合片区）
镇江内江滨水城市景观规划与设计

古典园林与建筑
上海东海观音寺建筑及环境设计
上海醉白池公园景观改造设计
扬中市佛教文化广场规划设计

道路绿化景观
泰州市长江大道景观设计
泰州溱湖大道景观改造设计
张家港金港大道景观提升设计
江阴霞客大道景观设计

农业科技园区
上海农科院奉浦院区规划、建筑、景观一体化设计（合作设计、景观施工）
天津市现代农业科技创新中心（西青）规划及建筑设计
天津市现代农业科技创新基地（武清）规划及建筑设计
鄱阳湖生态经济区现代农业科技创新示范基地规划及建筑设计
南昌市溪霞现代农业科技示范园规划、建筑、景观一体化设计
安徽省农业科学院（本部基地）规划设计

居住社区
上海青浦重固逸皓华庭景观规划设计工程(设计、施工一体化)
春申景城景观绿化工程（景观施工）
溧阳天目湖健康养生园景观规划设计
靖江新桥别墅区规划、建筑、景观一体化设计
博鳌亚洲论坛国际社区规划、建筑设计

亦境建筑景观

ADD：上海市普陀区中江路388号国盛中心1号楼3001-3003　（PC:200062）
　　　Rm. 3001-3003 Gouson Center Building 1#, 388 ZhongJiang Rd, Putuo District, Shanghai
TEL：021-6167 7866（总机）　　　　　　　　　FAX：021-6076 2388
http://www.edging.sh.cn

建筑设计研究院　　E-mail：arch@edging.sh.cn　　QQ：12 9321 8362
景观规划设计院　　E-mail：la@edging.sh.cn　　　QQ：19 5293 4808
景观工程公司　　　E-mail：gc@edging.sh.cn

注：数据均由上海亦境建筑景观有限公司统计

亦 小 亦 美 · 亦 真 亦 善

济南市园林规划设计研究院
JINAN LANDSCAPE PLANNING AND DESIGNING RESEARCH INSTITUTE

　　济南市园林规划设计研究院于1983年11月22日正式成立。拥有园林规划设计甲级、建筑设计乙级资质。院下设总工室、五个园林规划设计所、一个建筑设计所、效果图表现室、ST环境设计研究所（与日本景观设计大师德永哲先生合作成立）、济南彩叶园林工程项目管理有限公司等部门。我院所做设计项目获得部省市各级奖励100余项，其中国际奖项及国家级奖项16个，部级奖项7项，省级奖项24项。我院强调"虽由人做、宛自天开"源于自然而高于自然，把握传统园林的精髓，开阔思路，形成独具一格的设计风格。

传　真：0531-82970090
电　话：0531-82053043(办公室)
　　　　0531-82971782(经营部)
地　址：山东省济南市市中区六里山路10号
网　址：http://www.jnlad.com

Oasis 无锡绿洲
景观规划设计院有限公司

无锡绿洲景观规划设计院有限公司成立于2004年，具备建设部颁发的风景园林设计专项甲级资质。我们的宗旨是在不同的专业领域中，力求景观设计的功能性、创新性、人性化以及环保性。绿洲坚持运用当代设计手法及语言，将自然、人性与艺术作为不懈探索的设计命题，以务实的态度和高度的热情参与实践。

我们始终注重博采众长，不断创新，并且通过我们与客户之间的合作，建造可持续发展的环境。在城市公园、绿地及水系、风景旅游区、住宅及商业区等领域的规划设计中，提供了独特的解决方案和优秀的服务品质，得到了广泛的认可。

Landscape Design	Urban Planning	Architecture	Environment
景观设计	城市规划	建筑设计	环境咨询

WUXI LVZHOU

LANDSCAPE ARCHITECTURE DESIGN

Urban Planning
城市空间景观设计

Commercial/Business District Design
商业/办公区景观设计

Hotel Design
酒店景观设计

Residential Design
住宅区景观设计

School Design
学校景观设计

Attractions Design
生态公园景观设计

地址：江苏省无锡市滨湖区湖滨街 15 号鑫湖科技大厦 23 楼
电话：0510-66968096　0510-66968087
传真：0510-66968091　邮箱：lvzhou215@163.com
网址：www.wxlvzhou.com

二等奖 / THE SECOND PRIZE

以水为"纽带"链接城市间区域关系
——盘锦滨海新区景观规划设计
"CITY BOND" AMONG REGIONAL RELATIONSHIPS
—— LANDSCAPE PLANNING OF PANJIN COASTAL NEW AREA

中文标题：以水为"纽带"链接城市间区域关系——盘锦滨海新区景观规划设计
组　　别：本科
作　　者：贾晓丹
指导教师：李辰琦
学　　校：沈阳建筑大学
学科专业：景观建筑设计
研究方向：风景园林
分　　类：2013"园冶杯"风景园林（毕业作品、论文）国际竞赛规划类作品二等奖

Title: "City Bond" among Regional Relationships——Landscape Planning of Panjin Coastal New Area
Degree: Bachelor
Author: XiaodanJia
Instructor: Chenqi Li
University: Shenyang Architecture University
Specialized Subject: Landscape and Architecture Design
Topic: Landscape Architecture
Category: The Second Prize in Planning Group of the 2013 "Yuan Ye Award" International Landscape Architecture Graduation Project/Thesis Competition

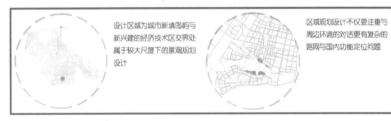

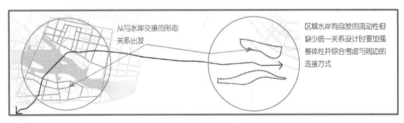

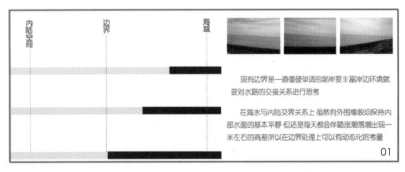

图01-02 基地现状与区域分析
Fig01-02 The basis of categorization and regional analysis

竞赛作品 CONTEST ENTRIES

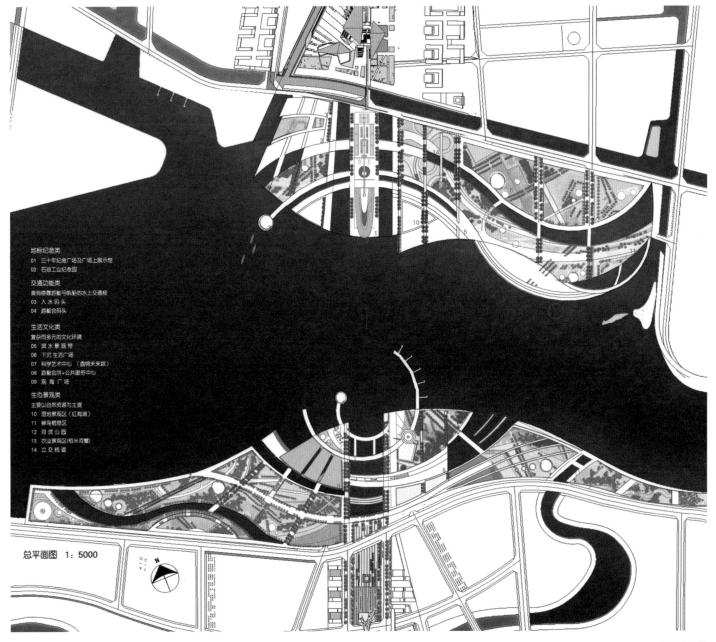

图 03 平面图
Fig03 Plan

作品简介：

水将陆分为两地也将陆连为一体如同一条纽带，"纽带"是指起到联系作用的东西。我的设计由此一衣带水的区域关系逐层展开。我始终坚信人是设计的本源，人的参与促发景观的最大价值。将人们的丰富活动和不同的行为体验融入到景观规划设计中去，才能创造出生态环境良好功能完善的滨水区域景观。

本次设计希望城市滨海边缘可以为快节奏的城市生活带来一丝轻松愉快的休闲体验。设置了多种参与性高的水上活动，公共空间开放性强，生态环境良好，希望为长时间生活在城市的人提供一个与日常生活不同的海滨体验（图01-02）。

设计将海水通过动感的流线引入城市，基地处于城市边缘是内陆城市规划的终点又与岛屿隔水相望。有起承转合的作用并可以形成多个重要的公共空间和自然景观。也因为基地的特殊地理位置，形成了以城市纽带为主题的方案。这不仅需要建立起基地与周边环境的良好对话，更需要在上层的统筹规划中使两岸关系形成统一的整体（图03，04，08）。

Introduction:

The "link" refers to the relation of things, and it means "linear" relationships among narrow strips of water areas in this project, so that the design is carried out layer by layer. I always believe that people are the source of design, and people's participation promotes the maximum value of landscape. Integrating different activities and behaviors of people into the landscape planning and design, we create a good ecological environment with a smooth functioning of the landscape waterfront.

Coastal edge design brings a pleasant and leisurely experience to the fast-pace of city life. By adding a variety of water activities with high participation, and setting public space and creating beautiful ecological environments, we hope to provide a different experience for people living in the city for a long period of time(Fig.01-02).

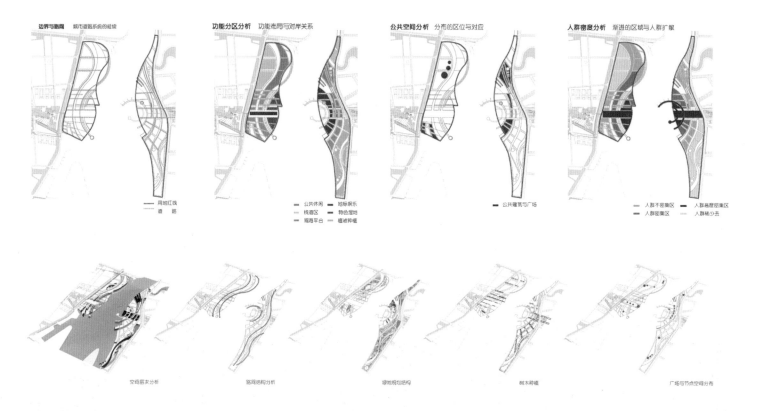

图04 方案系统分析
Fig04 System analysis

设计延续了城市既有的中轴线并结合空间序列在上半部以系统化的布局为主，中轴线节点为三十年纪念广场并设有城市地标。而下半部分以休闲娱乐为主，以虚化的水面空间与实体广场对应。整体的圆形关系体现了向心性使两岸更加整体和谐。

设计从城市结构的宏观角度出发 布局分析透视则以微观角度入手阐述空间的处理手法与人体尺度下的空间感受与体验使设计更加形象化。

设计中的圆形广场景观平台是有高差处理的，台阶与休息区绿化的结合方式在人体尺度下小空间的感受和体验变化方式形成良好小气候便于人的停留活动，易使人与人产生交流（图05）。

台阶的利用不仅可以有效控制岸线更能为人提供休闲平台，此次设计考虑到全年水位的变化，在边缘处理上让水位与台阶的高差相呼应产生一定的变化，人行其间有着良好的空间体验与视觉欣赏，在相对开阔的广场空间下有着细节的高差处理方式与亲水需求（图06）。

在核心交通区域的处理上以点作为停留，以线状空间作为穿行空间结合人群疏散广场进行布置以确保人流交通的通畅。将游艇俱乐部与公共服务中心放置于交通核上使人们能够快速到达为人的出行活动等需求提供方便快捷的路径选择才是真正做到设计理念中的以人为本（图07）。

The sea is brought into the city in dynamic flow. The site is located on the edge of the city at the limit of inland city planning, and across the water the island can be seen. Based on the special geographical location, the project takes "city link" as its theme. This not only the need to establish good communications with the surrounding environment, but also to form a unified whole of cross-strait relationship in the overall planning(Fig.03, 04, 08).

Continuing the systematic layout with the existing axis and spatial sequence in the upper area, we have located ThirtyYears Memorial Plaza and a city landmark at axial nodes. The lower area is positioned as the area of leisure entertainment with the virtual water space corresponding to the entity square. The circle embodies the idea that centrality helps create more harmonious relationships.

The design is based on the macro view of urban structure, while from the micro perspective, the layout analysis makes the design more pictorial by describing the spatial processing and the spatial feeling at a human scale. Through differences in elevation, in the design of the round landscape platform, we combine the footsteps and the greening in the rest area to create comfortable microclimates. This provides areas for residential activities and provides opportunities for people to communicate(Fig.05).

The utilization of footsteps can not only effectively control the coastline, but also provide leisure platforms for people. Given changes in the water level at different times of the year, we design the elevation of the steps based on the corresponding water level. An enjoyable well-functioning space with good views is provided through the processing the height differences and water levels in the relatively open square(Fig.06).

In the design of the core traffic area, we deal with the residences as points and the linear space as the travelling space, and ensure a smooth flow of traffic through the crowd evacuation square. The yacht club and the public service center are placed on the traffic core, in order to provide convenient path selection for people to reach their travelling destinations quickly. It is truly a people-oriented design concept(Fig.07).

查尔斯·沙（校订）
English reviewed by Charles Sands

局部节点透视 及其 设计分析

引水入城——创造一个连续并充满生气的城市边缘

加厚的边缘 叠加水岸
叠加空间 区分动静 形成供垂钓的水岸广场

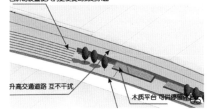

活力 动态化的景观
由水位变化产生的不同景观形象

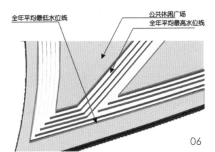

06

交通布局 与 联系方式
环网——明晰水上交通与陆路交通的联系方式

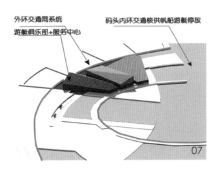

07

图 05-07 局部节点透视及其分析
Fig05-07 Analysis of the local node and its perspective

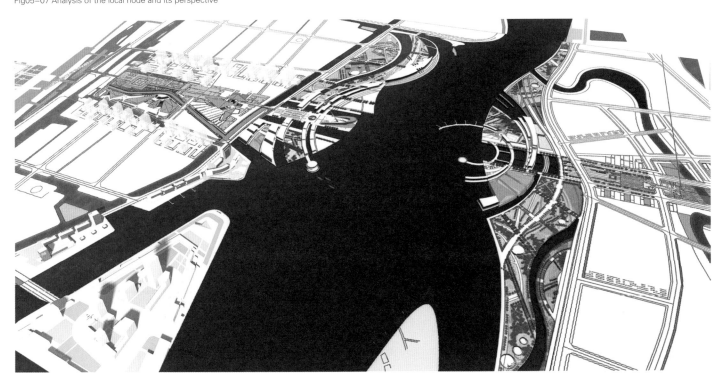

图 08 鸟瞰图
Fig08 Bird's eye view

衰落站区的新能量
——南京浦口火车站及其周边地区景观重组与整合设计

REGENERATION OF THE DISUSED STATION
——LANDSCAPE RECOMBINATION OF NANJING PUKOU RAILWAY STATION AND THE SURROUNDING AREAS

中文标题：衰落站区的新能量——南京浦口火车站及其周边地区景观重组与整合设计
组　　别：本科
作　　者：蒋沂玲　段倩　徐佳琪
指导教师：程云杉
学　　校：南京林业大学
学科专业：景观建筑设计
研究方向：园林规划
分　　类：2013"园冶杯"风景园林（毕业作品、论文）国际竞赛规划类作品二等奖

Title: Regeneration of the Disused Station——Landscape Recombination of Nanjing Pukou Railway Station and the Surrounding Areas
Degree: Bachelor
Author: Yiling Jiang, Qian Duan, Jiaqi Xu
Instructor: Yunshan Cheng
University: Nanjing Forestry University
Specialized Subject: Landscape and Architecture Design
Topic: Landscape Planning
Category: The Second Prize in Planning Group of the 2013 "Yuan Ye Award" International Landscape Architecture Graduation Project/Thesis Competition

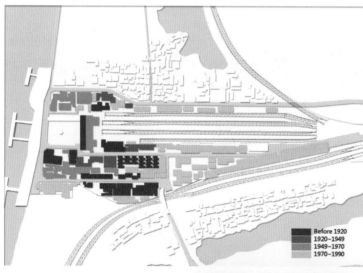

图 01 火车站区原貌
Fig01 The original appearance of the railway station district

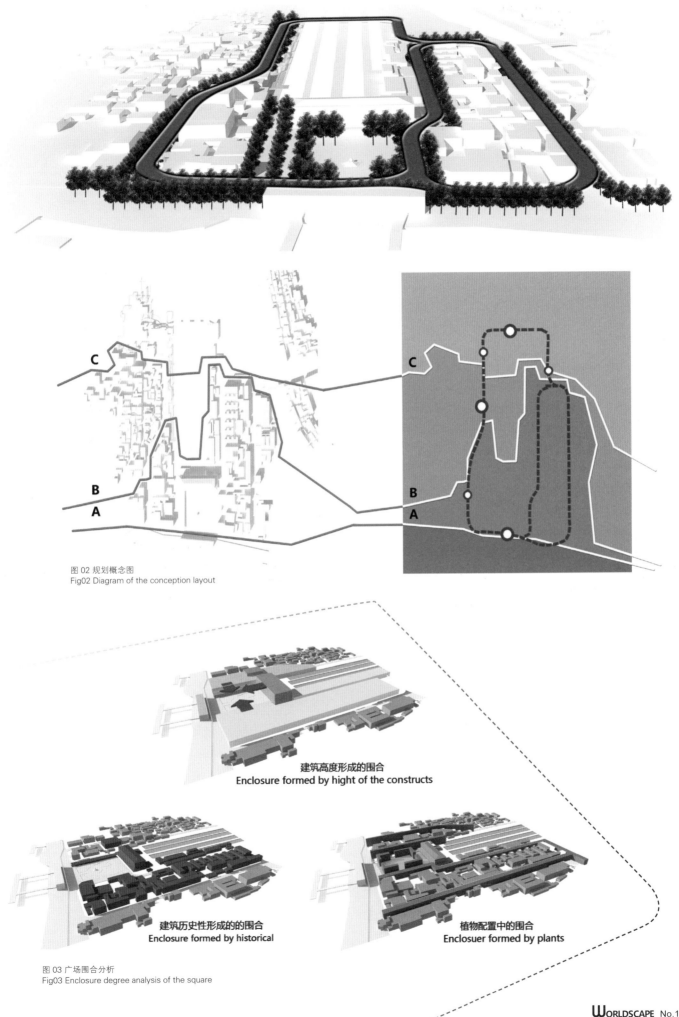

图 02 规划概念图
Fig02 Diagram of the conception layout

建筑高度形成的围合
Enclosure formed by hight of the constructs

建筑历史性形成的围合
Enclosure formed by historical

植物配置中的围合
Enclosuer formed by plants

图 03 广场围合分析
Fig03 Enclosure degree analysis of the square

图04 广场鸟瞰图
Fig04 Bird's eye view of the square

作品简介：

该地区是一处兼具民国建筑风貌和交通枢纽特征的城市空间，从激发场所多元价值入手（图01），将其分成三个区块（图02）。

首先，在最具历史价值的火车站广场区，遵循原有建筑风貌和空间尺度（图03），拓展广场可视空间，增强围合感，以突出其重要性（图04）。

其次，在有些陈旧的浦口老街部分，保护和修缮老建筑的同时，着力打造民国市井氛围，以促进特色商业街的形成（图05）。

而在更有发展空间的北部地区，规划了新的公园（图06）和广场（图07），以暗喻铁轨且变化多样的构架形式来承载丰富的现代生活（图08）和片片绿色（图09），并结合各类交通体系共同串接起差别显著但紧密关联的各个区块（图10）。

Introduction:

The site is comprised of an urban area that includes both the architectural style of Taiwan and the characteristics of a traffic hub (Fig.01). The area is divided into 3 blocks in order to motivate its multiple values (Fig.02).

Firstly, in order to show the importance of the train station square, which has the most important historical value, we broaden the square's visual space and enhance the sense of enclosure while keeping to the original architectural and spatial scale (Fig.03-04).

Secondly, in the old section of streets in Pukou we are trying to create a Taiwanese atmosphere to promote the formation of a commercial street while protecting and repairing the old structures (Fig.05).

The third section in the northerly direction has broad space for development so a new park (Fig.06) and Commercial Plaza has been designed here (Fig.07). The corridors in the park are metaphors for the train tracks, which are circuitous and fluctuant, carrying the richness of modern life (Fig.08-09). The transportations system links the three main sections, which are different but closely related (Fig.10).

图 05 老街及其绿化带
Fig05 Historical street and its green belt

图 06 公园廊架局部
Fig06 Part of the corridor in the park

图 07 商业区广场
Fig07 Square in front of the commercial section

图 09 攀援植物与公园廊架
Fig09 Corridor with vines

竞赛作品 CONTEST ENTRIES

图 08 室内休息室效果图
Fig08 Visual effect of the lounge

图 10 铁轨观光道局部图
Fig10 Sightseeing footpath adjoining the railway

查尔斯·沙（校订）
English reviewed by Charles Sands

土虱瓮 · 好挤好挤
——台中北屯区军功寮人文景观调查规划
TU SHR WENG · SIDE BY SIDE
——LANDSACPEING CONSERVATION PROJECT FOR THE CATFISH HUMANE COMMUNITY

中文标题：土虱瓮·好挤好挤——台中北屯区军功寮人文景观调查规划	Title: Landscaping Conservation Project for the Catfish Humane Community
组　　别：本科	Degree: Bachelor
作　　者：陈冠云、梁韦茵	Author: Guanyun Chen, Weiyin Liang
指导教师：戴宏一	Instructor: Hongyi Dai
学　　校：台湾勤益科技大学	University: National Chin-Yi University of Technology
学科专业：景观系	Specialized Subject: Landscape
研究方向：人文景观调查规划	Topic: Research and Design of Human Landscape
分　　类：2013"园冶杯"风景园林（毕业作品、论文）国际竞赛规划类作品二等奖	Category: The Second Prize in Planning Group of the 2013 "Yuan Ye Award" International Landscape Architecture Graduation Project/Thesis Competition

图 01 区域位置图
Fig 01 Location

图 02 土虱瓮巷弄景色
Fig02 View of tu shr weng alleys

图 03 综合分析
Fig03 Comprehensive analysis

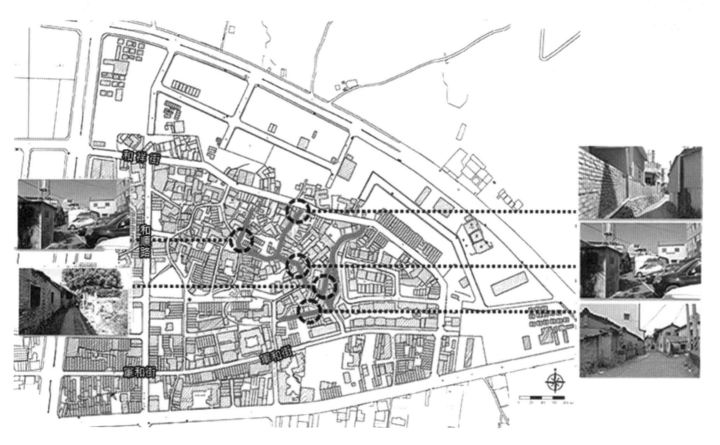

图 04 保留节点及建物
Fig04 Preservation construction

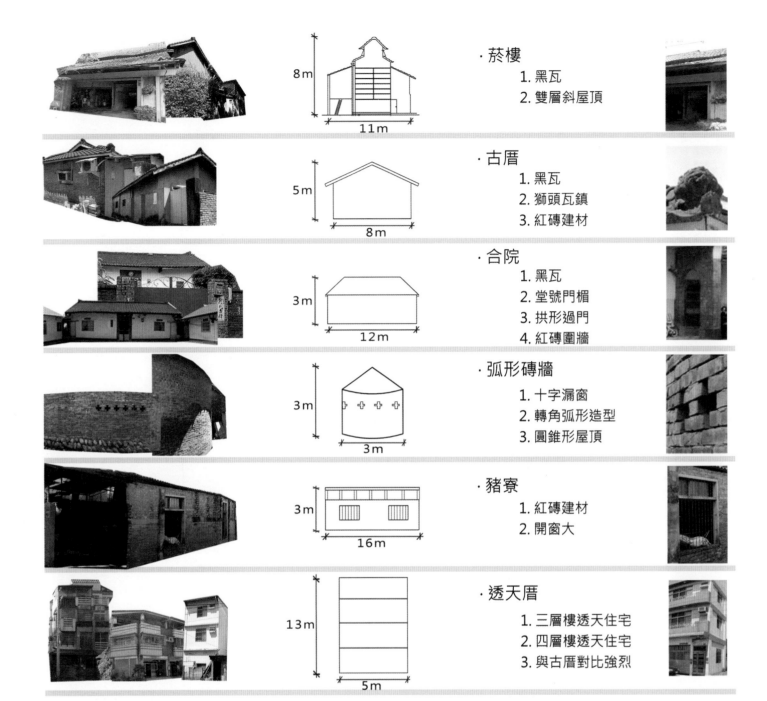

图 05 型态分析
Fig05 Pattern of construction analysis

作品简介：
台湾台中北屯区军功寮人文景观调查规划市北屯区境内的「军功寮」，是汉人在大坑地区最早开发的聚落（图 01）。军功寮庄内巷道曲折迂回，人口聚集稠密，宛如群居生活的「土虱」，此地区保存过去低矮短墙的合院，以及巷道曲折蜿蜒的建筑特色，故有「土虱厝」的称号（图 02-05）。

基地聚落周边发现军公墓文化遗址，其文化类型属于牛骂头文化，为特殊的文化历史遗迹，而当地区早期种植烟叶曾经辉煌腾达四十多年之久，当地拥有特色烟楼建筑。

本计划将藉由当地丰富的人文历史，保存当地独有的特色建筑，规划串连周边人文景点，形成具有传统特色文化的聚落空间（图 06-08）。

Introduction:
Jiun Gung Liau is a small town located in Beitun District of Taichung. This town was developed by the Han people. Jiun Gung Liau area has many long and narrow alleys (Fig 01). Due to the town's large population, the buildings were constructed side by side. Many of the older walled houses and courtyard houses, were protected and preserved by the residents of Jiun Gung Liau. Living in such close proximity was described as living like catfish, so this area was called Tu Shr Weng (Fig02-05).

In the past, the most dominant industry in the area was tabacoo farming. For over forty years the tobacco industry thrived. Nowadays, one of the most interesting aspect of the town is the specialty tobacco building.

Our team plans to preserve the uniqueness and traditional culture offered by Jiun Gung Liau area and prevent its destruction (Fig.06-08).

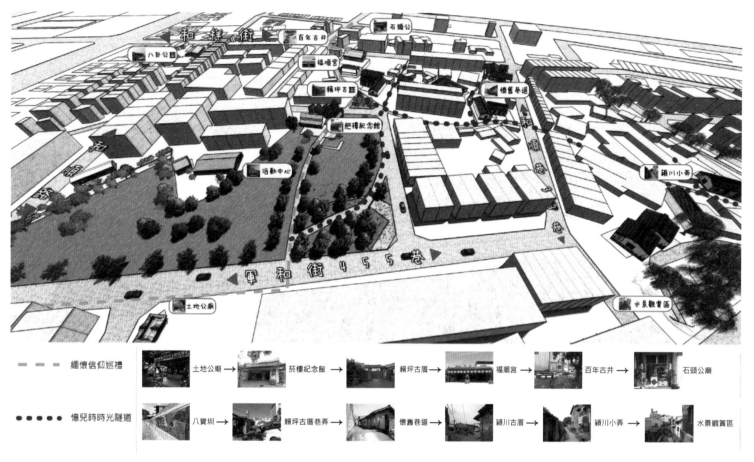

图 07 规划平面图
Fig07 Master plan

图 06 构想迭图
Fig06 Concept

图 08 都市设计准则
Fig08 The urban design guideline

查尔斯·沙（校订）
English reviewed by Charles Sands

美丽乡村，生态家园
——连云港市前云村改造规划设计
BEAUTIFUL COUNTRYSIDE, ECOLOGICAL HOMELAND
——PLANNING AND DESIGN OF QIANYUN VILLAGE IN LIANYUNGANG

中文标题：美丽乡村，生态家园——连云港市前云村改造规划设计
组　　别：本科
作　　者：唐菲　吴安琪　李旭彤
指导教师：郭苏明
学　　校：南京林业大学
学科专业：景观建筑设计
研究方向：景观建筑设计
分　　类：2013"园冶杯"风景园林（毕业作品、论文）国际竞赛规划类作品二等奖

Title: Beautiful Countryside, Ecological Homeland
—Planning and Design of Qianyun Village in Lianyungang
Degree: Bachelor
Author: Fei Tang, Anqi Wu, Xutong Li
Instructor: Suming Guo
University: Nanjing Forestry University
Specialized Subject: Landscape and Architecture Design
Topic: Landscape and Architecture Design
Category: The Second Prize in Planning Group of the 2013 "Yuan Ye Award" International Landscape Architecture Graduation Project/Thesis Competition

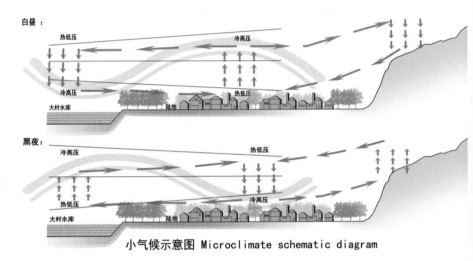

白天的风向是风从水库水面向村庄陆地吹，夜晚则相反。在前云村小气候环境中，水库起到了调节气温的作用，加强了空气流动，使气候更适合植物生长和人们居住生活。水库的存在优化了周边整体的环境质量。

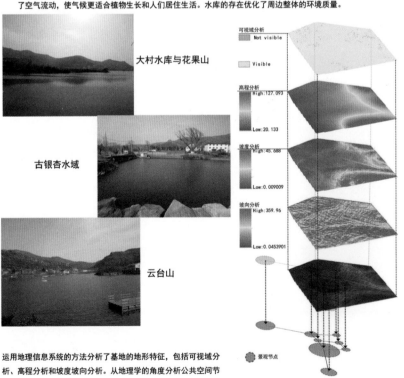

运用地理信息系统的方法分析了基地的地形特征，包括可视域分析、高程分析和坡度坡向分析。从地理学的角度分析公共空间节点的地形优势和景观优势。

图 01 小气候与 GIS 分析
Fig01 Microclimate and GIS analysis

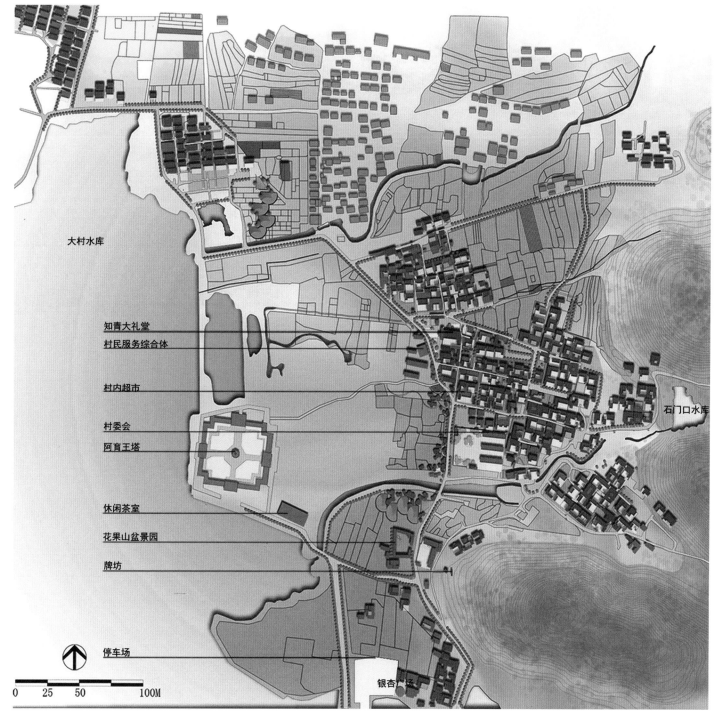

图 02 总平面图
Fig02 Master plan

作品简介：

以连云港花果山乡前云村为例探讨乡村景观规划与设计的方法。通过调研与查阅文献我们发现连云港花果山乡前云村乡村景观建设存在的问题，并分析其乡村景观资源特色（图 01）。结合景观生态学、可持续发展理论、风景园林规划设计学理论等多学科理论构建乡村景观规划与设计理论体系，最终提出一套适合连云港花果山乡前云村乡村景观建设的理论方法体系，并设计可行的改造方案，对于加强整治村庄环境综合水平，改善前云村人居环境起到积极的作用（图 02）。研究首先有助于改善前云村乡村景观的建设和发展，改变乡村景观建设千村一面、地域性淡化的现象，保护前云村乡村景观的整体性和地域性，使其恢复生机活力与吸引力；其次有助于协调前云村乡村景观资源的开发建设与自然生态环境保护之间的关系，促进乡村的可持续发展。课题的理论和方法可以丰富乡村景观规划与设计理论体系，并为我国其他类似地区的乡村景观建设提供参考借鉴（图 03-06）。

Introduction:

The project is based on the method of rural landscape planning and design in Qianyun village, Lianyungang Huaguoshan Township. Through research and literature review, we uncovered the problems of landscape construction in Qianyun village and analyzed its rural landscape resource characteristics (Fig.01). By Combining landscape ecology, sustainable development theory, landscape planning and design theory with other disciplines, we built a theoretical system of rural landscape planning and design. Based on this, we propose a suitable theoretical and methodological system of landscape construction in Qianyun Village, Lianyungang Huaguoshan Township. Under this system, a feasible rehabilitation programs is designed, which strengthens the integrated level of village environmental management and

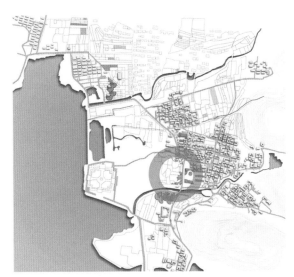
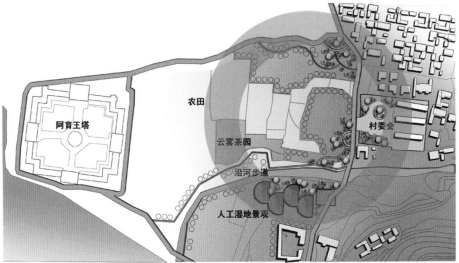

前云村村委会　　　云雾茶园　　　阿育王佛塔

图 03 村委会公共节点设计
Fig03 Design of nodes in village committee

竞赛作品 CONTEST ENTRIES

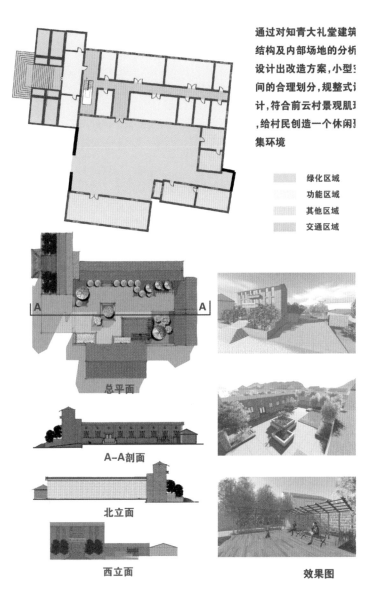

图 04 知青大礼堂环境改造
Fig04 Environmental reconstruction of educated youth auditorium

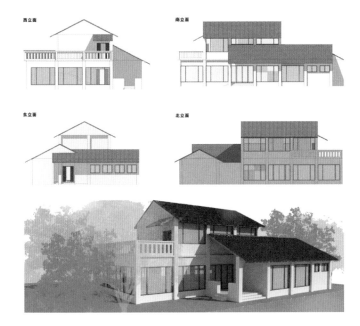

图 05 村民茶室设计
Fig05 Design of the teahouse

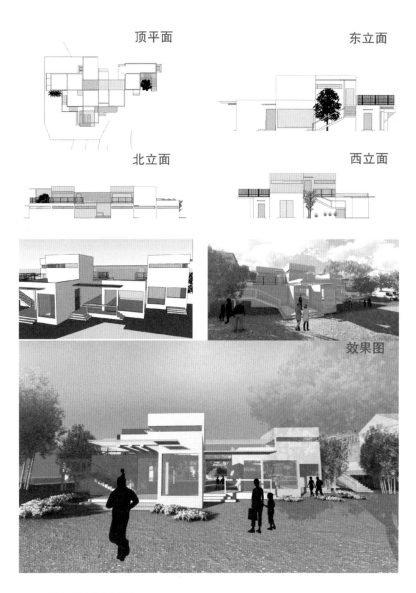

图 06 村民小型综合服务体设计
Fig06 Design of small integrated service system

improves the living environment of the Qianyun village (Fig.02). This research project helps to improve the landscape construction and development of the Qianyun village rural land and change the current situation whereby thousands of villages assume similar landscape characteristics and lose their regional features. This project contributes to restoring the vitality and appeal of the Qianyun village by protecting its territorial integrity. It also helps to coordinate the relationship between the development of Qianyun village's landscape resources and the protection of the natural ecological environment, and thus promotes sustainable rural development. Finally, the theoretical and methodological issues raised here can enrich the theoretical system of rural landscape planning and design, and provide references to the landscape construction of other similar areas in China (Fig.03-06).

查尔斯·沙（校订）
English reviewed by Charles Sands

"追寻记忆里的蝉噪与鸟鸣"
——上海嘉定嘉北郊野公园湿地涵养林景区设计
"PURSUING THE SONG OF CICADA AND BIRDS IN THE MEMORY"
—— THE DESIGN OF WETLAND CONSERVATION FOREST IN NORTHERN JIADING

中文标题:"追寻记忆里的蝉噪与鸟鸣"——上海嘉定嘉北郊野公园湿地涵养林景区设计
组　别:硕　博
作　者:黄川壑
指导教师:朱建宁
学　校:北京林业大学
学科专业:风景园林
研究方向:风景园林
分　类:2013"园冶杯"风景园林(毕业作品、论文)国际竞赛设计类作品二等奖

Title: "Pursuing the Song of Cicada and Birds in the Memory"
——The Design of Wetland Conservation Forest in Northern Jiading
Degree: Master
Author: Chuanhe Huang,
Instructor: Jianning Zhu
University: Beijing Forestry University
Specialized Subject: Landscape Architecture
Topic: Landscape Architecture
Category: The Second Prize in Design Group of the 2013 "Yuan Ye Award" International Landscape Architecture Graduation Project/Thesis Competition

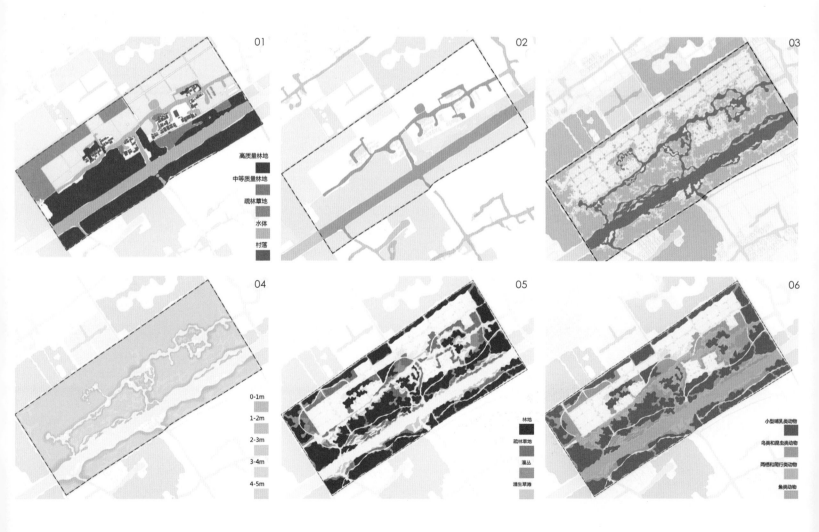

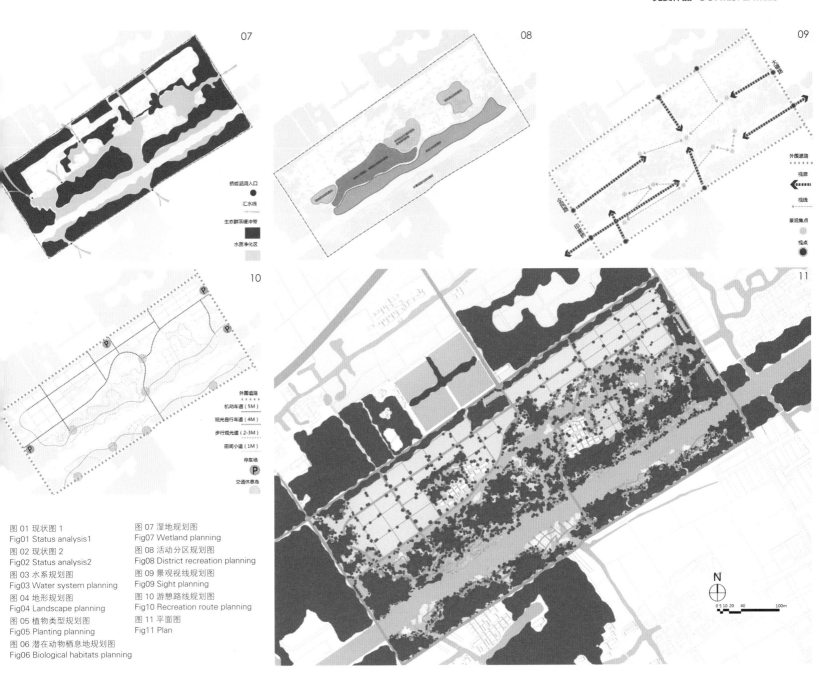

图 01 现状图 1
Fig01 Status analysis1

图 02 现状图 2
Fig02 Status analysis2

图 03 水系规划图
Fig03 Water system planning

图 04 地形规划图
Fig04 Landscape planning

图 05 植物类型规划图
Fig05 Planting planning

图 06 潜在动物栖息地规划图
Fig06 Biological habitats planning

图 07 湿地规划图
Fig07 Wetland planning

图 08 活动分区规划图
Fig08 District recreation planning

图 09 景观视线规划图
Fig09 Sight planning

图 10 游憩路线规划图
Fig10 Recreation route planning

图 11 平面图
Fig11 Plan

作品简介：

上海嘉北郊野公园湿地涵养林景区位于上海市嘉定区（方案平面图，方案鸟瞰图）。本设计首先考虑水系整理与水质净化。通过拓宽湖面，贯通整个水网，增加区域内水体总岸线的长度（图01－03），并通过小范围的挖方和填方营造各种类型的滩地，使间歇性淹水生境的种类丰富，营造多样的水岸环境（图04）。

其次是植被群落和动物栖息地的营建（图05－06）。通过引入乡土的湿生、中生植物种类，进行人为辅助下的湿地自然修复，建立稳定、高效、生物多样性丰富的生态系统。依据水位的变化种植相应的湿生群落，形成洼地、池塘、草地、滩涂和林地等多种生境类型。

根据不同的自然生境，在湿地景区内规划了北部、中部和南部3大区块。北部区块保留现有农田、鱼塘和村庄，经过调整成为生态农业休闲和观光旅游区，人工湿地较多。中部区块依托现有的水网和岛屿构筑沼泽湿地和河流湿地，并依据不同鸟类和小型动物的生活习性，

Introduction:

The Wetland forest will play a significant ecological role in the future(Master plan, Bird's eye view). We expect to build a healthy wetland system through scientific analysis and planning. The measures include:

(1) Establishing a wetland water system to demonstrate diversified water purification processes (Fig.01-04).

(2) Establishing a system of habitats aimed at attracting small animals and birds and enhancing biodiversity while establishing vegetation systems and creating diversified types of vegetation communities and habitats (Fig.05-06).

(3) Establishing a system of science education for people to explore riverbed wetlands, which also acts as a recreational system for socializing and relaxation (Fig.07-11,18).

传统游览观景体验

图12 设计概念
Fig12 Ideas

创造多样的栖息地，使之成为动物繁衍生息的家园。南部区块以湖泊湿地景观为主，广阔的水面，多样化水陆交接带和大片林地为各类生物的栖息提供了条件，也是市民开展休闲活动的场所，展现人与动物和谐相处的自然风光（图07-11,18）。

以人的观景视角为出发点，改变传统的站在平地上看景物的游览方式，在场地中结合原有地形加以高低起伏的地形，并创新性的利用"空中、地面、地下、水中"四个空间层次的栈道布置游线穿梭于整个场地中，引导人们近距离体验平日无法进入的空间维度（图12）。通过处在不同空间层次的栈道带领人们触摸半空中植物的树叶、花朵、果实，观看鸟类的巢穴（图13-15）；观察地面上美丽的灌木和小花小草，蚂蚁昆虫的活动，感受花朵的芬芳，触摸平静的水面；体会穿越水下空间的乐趣，观察水底鱼类的活动和水生植物的生长；观察地面动物的活动（图16-17）。以全新的视角观察景物，发掘新的视觉体验，并引发思考。

All of these factors contribute to the establishment of an attractive wetland park in northern Jiading. Based on the existing conditions, the Wetland district is designed differently from north to south. The paddy field is placed in the north zone to form an agricultural tourism experience. In the central zone, two types of wetlands are designed—surface flow wetlands and undercurrent wetland—to improve water quality and regulate water levels. In the south, due the dense river network and thick plant growth, we propose diversified water features and abundant vegetation communities to provide a breeding place for birds and animals. The south zone is an open water area. In this area, an array of islands are designed with different ecological roles, such as bird habitats, and water purification demonstration areas. A discovery trail is established for visitors to experience the natural succession process of the wetlands. At the same time, the seasonal wetland presents different landscapes in the rainy and dry season.

Our design takes people's viewpoint as a starting point, with the hope of changing ingrained prevailing attitudes. The undulating terrain is designed with the innovative use of pipelines moving through the various terrain types. Visitors are taken on a journey through four themed spatial levels, which include: the air, the ground, underground, and underwater. With this new perspective, visitors can not only explore new visual experiences, but also learn and enjoy nature (Fig.12). By arranging a suitable route, we hope to encourage people to touch the leaves, flowers and fruit and observe shrubs and flowers as well as the activities of ants and insects on the ground (Fig.13-15). There will be opportunities to smell the fragrance of flowers and touch calm water; to see fish, and aquatic plants when crossing through the underwater space; to perceive the strong roots of plants and underground animals and experience the darkness of the underground space (Fig.16-17), all of which are so foreign to the human living environment.

查尔斯·沙（校订）
English reviewed by Charles Sands

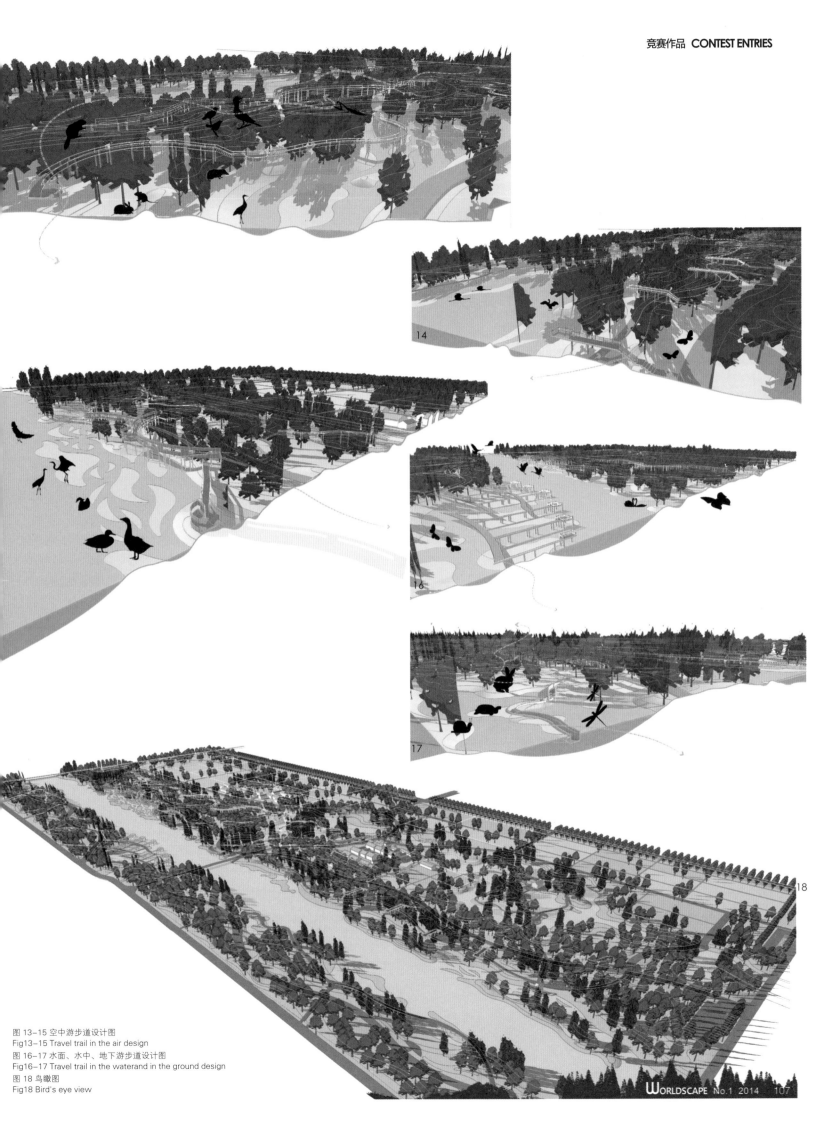

图 13-15 空中游步道设计图
Fig13-15 Travel trail in the air design
图 16-17 水面、水中、地下游步道设计图
Fig16-17 Travel trail in the waterand in the ground design
图 18 鸟瞰图
Fig18 Bird's eye view

金三角绿洲
——汕头市北郊公园规划设计
THE GOLDEN TRIANGLE OASIS
——NORTHERN SUBURB PARK PLANNING AND DESIGN IN SHANTOU

中文标题：金三角绿洲——汕头市北郊公园规划设计
组　　别：硕博
作　　者：陈磊晶　郑　懿　肖红涛
指导教师：李　敏　罗红梅
学　　校：华南农业大学
学科专业：风景园林规划设计
研究方向：景观规划设计
分　　类：2013"园冶杯"风景园林（毕业作品、论文）国际竞赛设计类作品二等奖

Title: The Golden Triangle Oasis——Northern Suburb Park Planning and Design in Shantou
Degree: Master
Author: Leijing Chen, Yi Zheng, Hongtao Xiao
Instructor: Min Li, Hongmei Luo
University: South China Agricultural University
Specialized Subject: Landscape Planning and Design
Topic: Landscape Planning and Design
Category: The Second Prize in Design Group of the 2013 "Yuan Ye Award" International Landscape Architecture Graduation Project/Thesis Competition

图01 鸟瞰图
Fig01 Bird's eye view

01 大学路公园主入口
   Main entrance
02 游船码头
   Boat wharf
03 高架景观廊道
   The elevated corridor
04 钻石结婚礼堂
   The diamond wedding center
05 儿童戏水池
   Children swimming center
06 尘世舞台
   City stage
07 香草摄影天地
   Flower shooting area
08 服务建筑
   Service buildings
09 潮汕路公园主入口
   Second main entrance
10 沙地门球场
   Sand gate
11 盆景文化园
   Potted culture garden
12 尘世雕塑
   City sculpture
13 金凤路公园主入口
   Third main entrance
14 小型动物园
   Zoo
15 体育中心
   Sports center
16 儿童拓展营
   Children's camp
17 儿童拓展营入口拓展营
   Children's camp
18 公园主干道及单车绿道
   Main road & green way

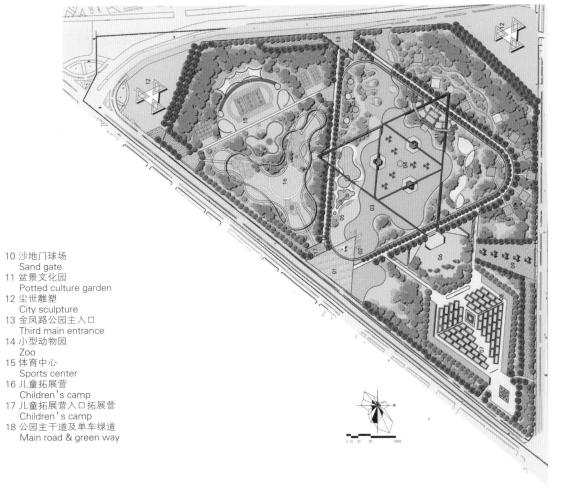

图02 总平面图
Fig02 Master plan

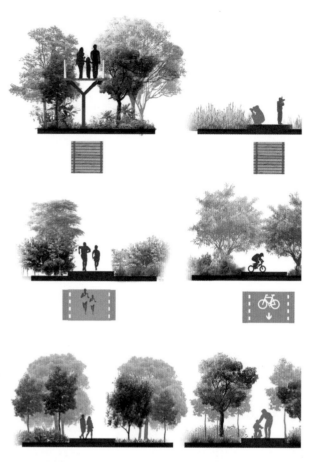

图03 道路类型分析
Fig03 Path analysis

作品简介：

本项目场地位于汕头市区西北交通枢纽地带，由潮汕路、大学路和金凤路三条市区主干道围合成的"三角形"地块，规划构思的来源于潮汕平原的大"三角"地理特征，与项目地块的小"三角"特征相结合。我们运用"三角形"构图，合理搭配几何图形，将"三角形"提升，塑造景观的核心和标志——"钻石"，打造金三角绿洲（图01-02）。本方案结合多元的潮汕文化与创新的南洋文化，营造多元的生活体验（图03-07），创新的乐活感受，形成社会与自然完全结合的城市生态。

Introduction:

The site of this project is located in the traffic hub region in northwest Shantou. It is a trianglular plot enclosed by three main urban roads, Chaoshan Road, University Road and Jinfeng Road. The planning idea comes from the combination of the "Big Triangle" geographical features of Chaoshan plain and the "Small Triangle" of the plot. Collocating with appropriate geometric figures, we apply the "triangle" composition during the design and abstract the image of a "Diamond". This shape represents the core and the symbol of the the Golden Triangle Oasis (Fig.01-02). By combining diverse Chaoshan culture with innovative Nanyang culture, this design creates pluralistic life experiences (Fig.03-07) and a fresh LOHAS feeling. It, thus contributes to form an urban ecology that fully combines society with nature .

图04 高架廊道效果图
Fig04 Elevated road rendering

图 05 钻石结婚礼堂
Fig05 Diamond wedding hall

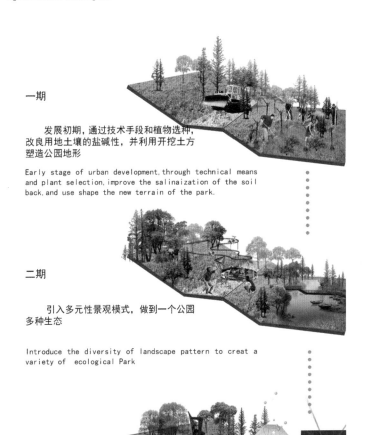

一期

发展初期，通过技术手段和植物选种，改良用地土壤的盐碱性，并利用开挖土方塑造公园地形

Early stage of urban development, through technical means and plant selection, improve the salinaization of the soil back, and use shape the new terrain of the park.

二期

引入多元性景观模式，做到一个公园多种生态

Introduce the diversity of landscape pattern to creat a variety of ecological Park

三期

随着商业、居住以及文化三者趋于平衡并焕发活力，将引入丰富多彩的体验景观

With commercial, residential and cultural balance and radiate strong vitality, introduce colorful pratice.

图 06 公园发展模式
Fig06 Development mode

图 07 香草广场效果图
Fig07 Square rendeing

查尔斯·沙（校订）
English reviewed by Charles Sands

图 01 鸟瞰图
Fig01 Bird's eye view

微山微水微声
——李光地读书处生态文化景观设计
MICRO HILL, MICRO WATER, MICRO SOUND
——THE ECOLOGICAL CULTURAL LANDSCAPE DESIGN OF LIGUANGDI'S READING PLACE

中文标题：微山微水微声—李光地读书处生态文化景观设计
组　别：本科
作　者：侯喆昊
指导教师：张　恒　桑晓磊
学　校：华侨大学
学科专业：建筑与城市环境艺术设计
研究方向：景观设计
分　类：2013"园冶杯"风景园林（毕业作品、论文）
　　　　国际竞赛设计类作品二等奖

Title: Micro Hill, Micro Water, Micro Sound—The Ecological Cultural Landscape Design of Liguangdi's Reading Place
Degree: Bachelor
Author: Zhehao Hou
Instructor: Heng Zhang, Xiaolei Sang
University: Huaqiao University
Specialized Subject: Design of Architecture and Urban Environmental Art
Topic: Landscape Design
Category: The Second Prize in Design Group of the 2013 "Yuan Ye Award" International Landscape Architecture Graduation Project/Thesis Competition

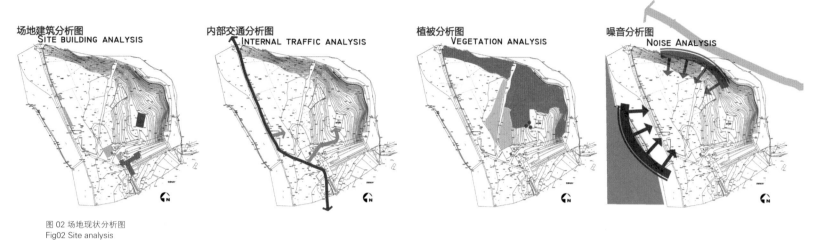

图02 场地现状分析图
Fig02 Site analysis

作品简介：

该基地位于泉州市湖头镇北部工业片区内（图01）。场地内部条件复杂，自然山体与湿地围合而成的自然空间形态面临着周边噪音的影响（图02）。本设计力图通过景观手段解决场地问题，从而达到微山微水微声的景观状态。同时场地也是李光地少年读书处遗址，继承传统文化，给游客带来读书的文化科普教育。并将场地废弃的电冶厂重新激活与更新，场地体现着现代工业与传统闽南文化的交融与碰撞，形成后工业文化景观，既是对传统李光地读书文化的追溯又是对工业文明记忆的传承。营造服务片区居民与城镇居民娱乐休闲与科普体验的城镇生态文化公园（图03-07）。

Introduction:

The site is located in the northern industrial area, Lake Town, Quanzhou City (Fig.01). Apart from complex internal conditions, the natural space is comprised of an enclosure of mountains and wetlands, which face the influence of the surrounding urban noise (Fig.02). We are trying to solve the problem through an approach to landscape design that aims for the favorable state of: Micro hill, Micro water and Micro sound. Because the site was the reading spot for Guangdi Li in his youth, as well as reflecting the traditional culture, it also puts forward the educational theme of reading for tourists. By "reactivating" and "renewing" the abandoned electrometallurgy plant, which represents the mergence and collision between modern industry and traditional southern Fujian culture, we create a landscape of post-industrial culture. It is a retrospective of traditional reading culture as well as a continuation of industrial memory. In a word, this design mainly focuses on creating a cultural eco-park for residents' recreational activities and historical education (Fig.03-07).

图03 剖面图
Fig03 Profiles

图 04 入口广场设计
Fig04 Design of entrance square

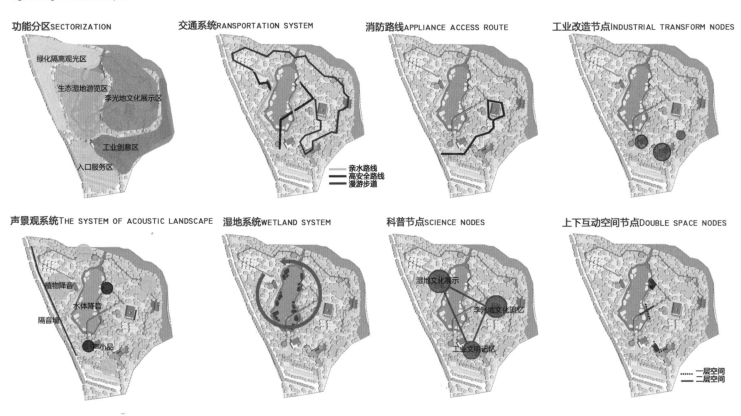

图 05 设计诠释
Fig05 Design interpretations

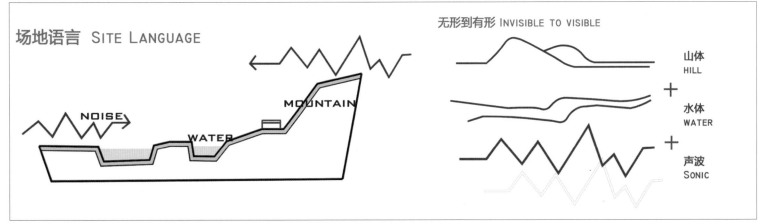

图 06 设计语言
Fig06 Design language

竞赛作品 CONTEST ENTRIES

图 07 平面图
Fig07 Plan

查尔斯·沙（校订）
English reviewed by Charles Sands

针灸术
——城市绿楔的后棕地生态修复
ACUPUNCTURE
——THE ECO-REMEDIATION OF POST-BROWNFILEDS IN A WEDGE-SHAPED GREEN SPACE

中文标题：针灸术——城市绿楔的后棕地生态修复	Title: Acupuncture——The Eco-Remediation of Post-Brownfileds in a Wedge-shaped Green Space
组　　别：本科	Degree: Bachelor
作　　者：张　姝　杨文祺	Author: Zhu Zhang, Wenqi Yang
指导教师：熊和平	Instructor: Heping Xiong
学　　校：华中科技大学	University: Huazhong University of Science and Technology
学科专业：景观学	Specialized Subject: Landscape Science
研究方向：风景园林设计	Topic: Landscape Architecture Design
分　　类：2013"园冶杯"风景园林（毕业作品、论文）国际竞赛设计类作品二等奖	Category: The Second Prize in Design Group of the 2013 "Yuan Ye Award" International Landscape Architecture Graduation Project/Thesis Competition

作品介绍：

场地位于安徽省六安市淠北新区的城市近郊（图01-03）。总规划面积10.23km²，建设周期十年，预将其打造为一个集生态、文化、游憩为一体的山体郊野公园。山体由城市弃土堆砌而成，针对现状我们采用"堆放弃土和绿化同时进行"以及"针灸式生态修复"的技术手段，使其成为城市中轴线的端点及制高点（图04-05）。

1 堆放弃土和绿化同时进行：

为期十年的公园建设和堆山工程，将会导致水土流失、沙尘漫天等问题。采用层层回填，层层绿化的方式解决问题（图06）。

2 针灸式生态修复：

在合适的位置栽植"林点"，林点通过自然力传播种子形成林片，林片相对较稳定扩散至整个公园。和中医的针灸法一样，我们试图通过树林的打点方式以最少的资金和人力修复整个场地的生态（图07-10）。

Introduction:

The site is located in the suburb of Pi Bei new-development district of Liu An City, Anhui Province (Fig.01-03). The total planning area is 10.23 square kilometers. The construction is scheduled to be completed over ten years. This design aims to build a mountain country park, which integrates ecology, culture and recreation. The mountains on the site will require extensive geo-engineering, therefore we apply the strategy of "layering earth while greening" and "ecological restoration in acupuncture style". The site sits at the end of the central urban axis at a commanding height over the city (Fig.04-05).

1. Layering earth while greening:

The ten-year project of park construction and mountain engineering work will lead to soil erosion and dust pollution. The problem can be solved through a method of backfill and greening layer upon layer (Fig.06).

2. Ecological restoration in acupuncture style:

Firstly, we plant "green points" in suitable locations, and then the "green points" can gradually disseminate seeds, which will naturally grow to form "green patches". Eventually these "green patches" will spread over the entire surface of the park. Like the acupuncture of traditional Chinese medicine, we are trying to restore the ecological environment of the site with minimum capital and manpower through the strategy of green-points (Fig.07-10).

图01 总平面图
Fig01 Master plan

图02 现场照片
Fig02 Site photos

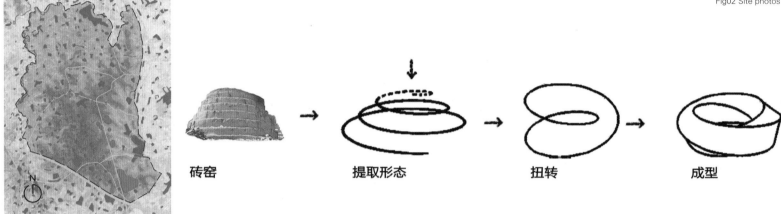

现状道路
现状水体
建筑弃土堆放

图03 现场问题
Fig03 Site problems

砖窑 → 提取形态 → 扭转 → 成型

图04 建筑生成
Fig04 Architecture generation

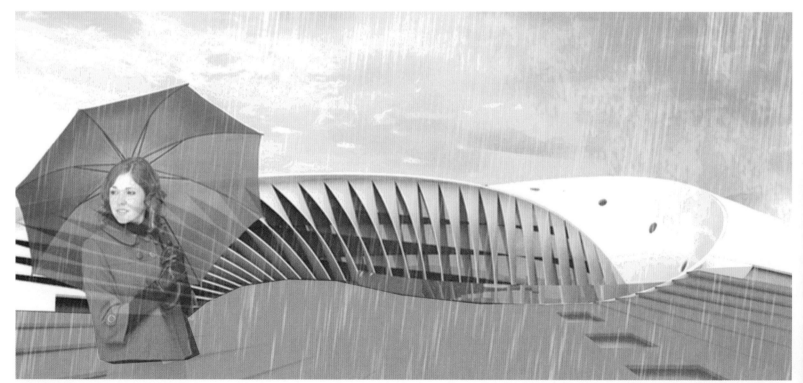

图05 建筑效果图
Fig05 Architecture rendering

竞赛作品 CONTEST ENTRIES

图06 堆土绿化技术
Fig06 Design of entrance square

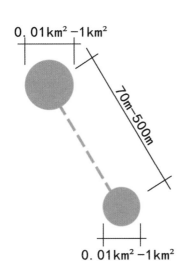

图07 针灸点
Fig07 Acupuncture points

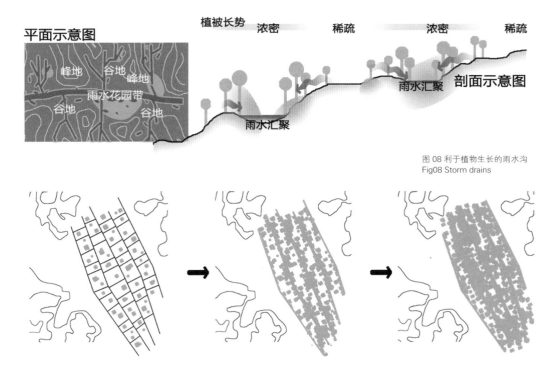

图08 利于植物生长的雨水沟
Fig08 Storm drains

图09 林点的蔓延
Fig09 Green spreading

图10 效果图
Fig10 Master rendering

查尔斯·沙（校订）
English reviewed by Charles Sands

铁路公社
——南京浦口火车站改造设计分析
RAILWAY COMMUNE
——DESIGN AND ANALYSIS OF NANJING PUKOU RAILWAY STATION

中文标题：铁路公社——南京浦口火车站改造设计分析
组　　别：本科
作　　者：高枫　车建安　刘熠　王辉
指导教师：张哲
学　　校：南京林业大学
学科专业：景观建筑设计
研究方向：风景园林、景观建筑设计
分　　类：2013"园冶杯"风景园林（毕业作品、论文）
　　　　　国际竞赛设计类作品二等奖

Title: Railway Commune——Design and Analysis of NanJing Pukou Railway Station
Degree: Bachelor
Author: Feng Gao, Jian'an Che, Yi Liu, Hui Wang
Instructor: Zhe Zhang
University: Nanjing Forestry University
Specialized Subject: Landscape and Architecture Design
Topic: Landscape Architecture, Landscape and Architecture Design
Category: The Second Prize in Design Group of the 2013 "Yuan Ye Award" International Landscape Architecture Graduation Project/Thesis Competition

① 时尚铁轨T台
② 旧厂房改造
③ 红色平桥
④ 趣味车厢宿舍
⑤ 火车历史博物馆
⑥ 特色铁轨景观

图01 总平面图
Fig01 Master plan

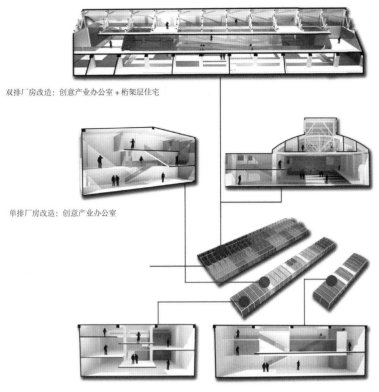

双排厂房改造：创意产业办公室＋桁架层住宅

单排厂房改造：创意产业办公室

将三个大厂房分别按照大中小型企业划分，对于不同大小的空间，选取典型案例进行空间利用的探索。

图 02 旧厂房改造
Fig02 Old factory building transformation

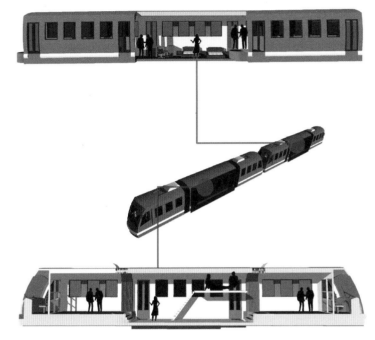

图 03 趣味车厢宿舍
Fig03 Train carriage dormitories

作品简介：

浦口火车站作为城市发展、更新换代的遗留见证者，是众多存在于城市的历史遗留，未被合理再利用的尴尬地之一。如何将弃用的火车站结合自身的特色变废为宝，是解决"尴尬"的重要突破点。而最终的"铁路公社"这一概念的提出（图01），则是充分考虑到周边居民的景观需求，社会群体的怀旧需求，创业人士的就业需求以及艺术家、设计师们的审美需求，运用旧厂房功能再改造（图02）、趣味车厢宿舍（图03）、时尚铁轨T台（图04）、火车历史博物馆（图05）以及特色铁轨景观（图06）等元素，打造出别具一格，极具地区特色的铁路公社，变废为宝，为遗留历史景观地段免除了弃之可惜、用之无地的尴尬。铁路公社以满足不同人群需要出发，充分利用自身的地理优势，势必成功吸引众多人群前来参观，游玩，创业和休憩（图07–10）。

Introduction:

A witness to the city's development and transformation, Pukou railway station has become an important historical construction without any obvious strategy of reuse. The key is finding a way to turn an "abandoned railway station" into something of value by making use of its own characteristics.

Given the wide-ranging requirements of the project relating to the landscape, the surrounding residents, historical preservation, employment, and aesthetics, we came up with the concept of "Railway Commune" (Fig.01). We utilize and reference many existing elements, for example, a functional reconstruction of the old factory (Fig.02), dormitories designed from train carriages (Fig.03), a fashion runway track (Fig.04), a railroad history museum (Fig.05) and a special railway themed landscape (Fig.06), to create a railway commune with a unique regional characteristics. Waste is turned into treasure and the problem of historical structure reuse is solved. Taking different kinds of requirements into consideration and making full use of its own regional advantages, railway commune will attract people to visit, play, work and rest (Fig.07-10).

图 04 时尚铁轨 T 台
Fig04 Fashion runway track

图 05 火车历史博物馆
Fig05 Train historical museum

图 06 特色铁轨景观
Fig06 Special railway landscape

竞赛作品 CONTEST ENTRIES

图 07 鸟瞰图
Fig07 Bird's eye view

图 08 节点 1
Fig08 Node one

图 09 节点 2
Fig09 Node two

图 10 节点 3
Fig10 Node three

查尔斯·沙（校订）
English reviewed by Charles Sands

HUAYU 华宇园林

住建部园林工程施工一级企业
设计.施工.苗木.养护

HUAYU 华宇园林

重庆华宇园林股份有限公司
CHONGQING HUAYU
LANDSCAPE&ARCHITECTURE CO.,LTD.

■ 全国用户满意企业称号　■ 重庆市优秀民营企业　■ 重庆市著名商标

ADD:重庆市江北区北滨一路506号　TEL:0086-023-63670488 FAX:0086-023-63670488
PC:400021 E-MAIL:HUAYUYUANLIN@126.COM
设计分公司　TEL:67999312　FAX:67999311
www.cnhyyl.com

北京丽宫

守于心 道于行

www.aandi.net

A&I INTERNATIONAL
安道国际

源树景观（R-land）是国内顶级的专业 环境设计机构。自2004年成立以来，通过不懈的努力，在景观规划、公共空间、旅游度假、主题设计等领域都获得了傲人的成绩，特别是在高端地产景观的咨询及设计方面，处于绝对的领先地位。

　　源树景观（R-land）的设计团队中汇集了大量的国内外景观设计精英，其主要设计人员都曾在国内外高水平设计单位中担任重要职务，严格的设计流程确保了每一项设计作品的完美呈现。

　　源树景观（R-land）历经数年，已完成了数百项设计任务，其中：河北省邯郸市赵王城遗址公园、中关村创新园、山东荣成国家湿地、西安大唐不夜城、北京汽车博物馆、龙湖"滟澜山"、天津团泊湖庭院、招商嘉铭珑原、远洋傲北、中建红杉溪谷、西山壹号院等若干项目均已建成并得到各界的广泛认可。

　　源树景观致力于最高品质的景观营造，力求为合作方提供最高水准的设计保障。

Add: 北京市 朝阳区朝外大街怡景园 5-9B（100020）　Tel:（86）10-85626992/3　85625520/30　Fax:（86）10-85626992/3　85625520/30 - 5555　Http://www.r-land.com

R-land

Beijing -Tianjin -Tokyo -Sydney　　YS Landscape Design

http://www.r-land.cn　源树景观

景观规划 Landscape Planning　　公共空间 Public Space　　居住环境 Living environment　　主题设计 Theme Design

设计类最佳人文奖 / THE BEST HUMANIST PRIZE OF DESIGN GROUP

沈阳薰衣草庄园景观设计
LAVENDER MANOR LANDSCAPE DESIGN, SHENYANG

中文标题：沈阳薰衣草庄园景观设计
组　别：本科
作　者：陆书雯
指导教师：孙振邦
学　校：沈阳建筑大学
学科专业：景观建筑设计
研究方向：自然生态 人文 低碳景观
分　类：2013"园冶杯"风景园林（毕业作品、论文）
　　　　国际竞赛设计类作品最佳人文奖

Title: Lavender Manor Landscape Design, Shenyang
Degree: Bachelor
Author: Shuwen Lu
Instructor: Zhenbang Sun
University: Shenyang Architecture University
Specialized Subject: Landscape and Architecture Design
Topic: Natural Ecology, Humanity, Low-carbon Landscape
Category: The Best Humanist Prize of Design Group in the 2013 "Yuan Ye Award" International Landscape Architecture Graduation Project/Thesis Competition

图01 基地现状图1
Fig01 Base map 1

Q：现在给你一个机会，让你重新设计这里的户外空间，你会增加什么？你会删除什么？你会改变什么？

花园式户外空间，增加户外休息空间，供成年人和孩子们玩耍之余进行休息，使种植花卉区域与树木种植相结合，能够天热躲避太阳，将古典主义与北欧文化相结合，将感性与理性相结合，以原生态为基础，贯穿整体设计，加点有特色的景观，进行有韵律的整体规划

同样的问题，不一样的回答，表达同一种需求

但是
现在
什么也没有……

图 02 基地现状图 2
Fig02 Base map 2

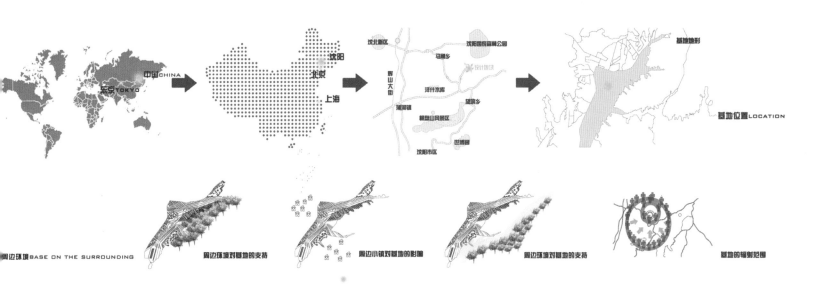

图 03 基地周边环境分析
Fig03 Analysis of surrounding environment

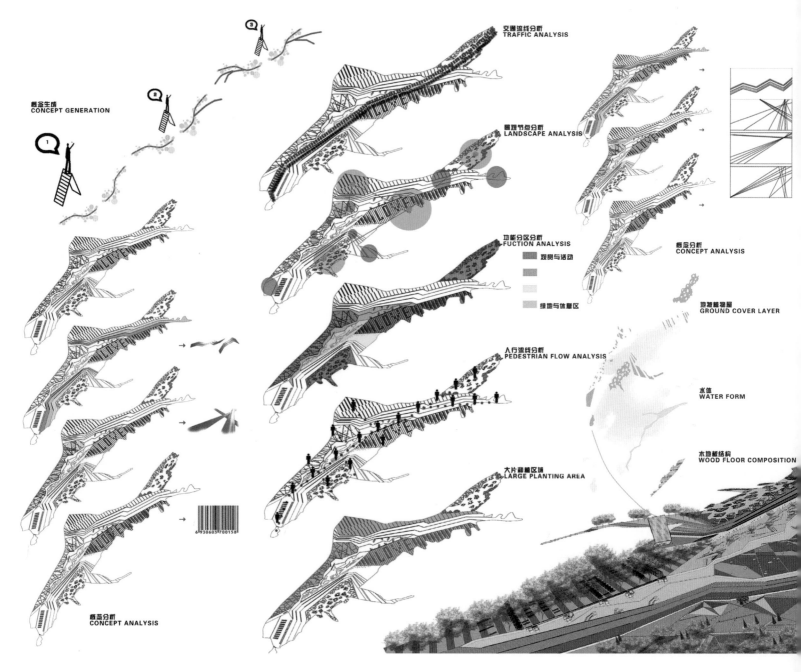

图04 分析图
Fig04 Analysis diagram

作品简介：

人类赖以生存的地球面临着十分紧急的情况，60%的生态系统在过去的数十年间遭受到严重的破坏，预计在2050年物种灭绝的速度要比自然淘汰的速度快100-1000倍。我们必须跟时间赛跑，以尽快停止这个毁灭性的过程。我们希望以自然利益为重，不再与自然对抗，而是同自然共同协调发展（图01-04）。

不论着手于小处还是全球尺度，人们都必须要优先考虑保护生物的多样性，因为这是维持生态系统平衡的根源所在，也是生态抵抗外界冲击的有力工具。多样性的生态环境不仅能提供生命所需的氧气、滋养动植物的生长，也对调节气候起到非常重要的作用（图05-09）。

Introduction:

Now we are faced with the ecological crisis of 60% of the ecosystem suffering from enormous damage inflicted over recent decades while the rate of extinction of endangered species is predicted to be 100 to 1000 times higher than the natural cycle by 2050. Indisputably, it is a race against time. People must commit to stopping this destructive process, by helping and cooperating with nature, even when it is detrimental to short-term economic profit(Fig.01-04).

Therefore priority must be given to the biodiversity of all living beings, from the minutest scale to the scale of the planet. The development of biology diversity is the primary means of balancing the ecosystem, and the inherent facilitator of ecological resilience. A biologically diverse environment can provide vital oxygen and nourish flora and fauna, and thus contribute to maintaining the natural cycles such as water and climate (Fig.05-09).

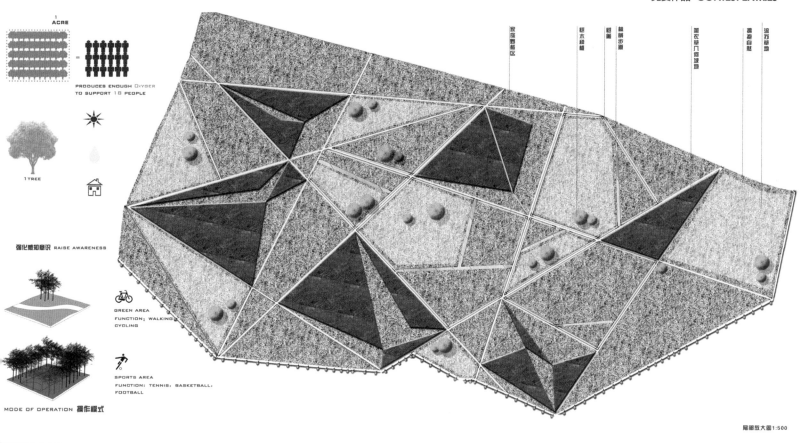

图05 节点放大图
Fig05 Node enlargement

图06 浪漫表白园地效果图
Fig06 Renderings of romantic garden

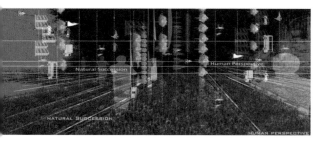

图07 薰衣草种植体验区效果图
Fig07 Rendering of lavender planting experience area

图08 家庭露营用地效果图
Fig08 Rendering of family camping area

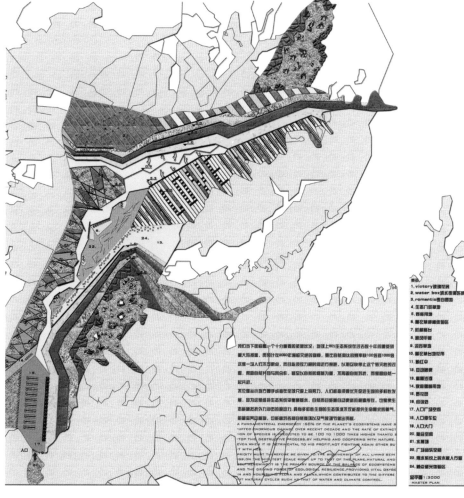

图09 总平面图
Fig09 Master plan

查尔斯·沙(校订)
English reviewed by Charles Sands

图01 生活现状
Fig01 Life status

低碳、低术、低生活
——农民工工地生活空间景观策略
LOW CARBON, LOW TECHNOLOGY, LOW IMPACT
——MIGRANT WORKER RESIDENTIAL LANDSCAPE STRATEGY

中文标题：低碳、低术、低生活 ——农民工工地生活空间景观策略
组　　别：本科
作　　者：高东东
指导教师：郑洪乐
学　　校：福建农林大学
学科专业：风景园林
研究方向：低碳景观
分　　类：2013"园冶杯"风景园林（毕业作品、论文）国际竞赛设计类作品最佳人文奖

Title: Low Carbon, Low Technology, Low Impact——Migrant Worker Residential Landscape Strategy
Degree: Bachelor
Author: Dongdong Gao
Instructor: Hongle Zheng
University: Fujian Agriculture and Forestry University
Specialized Subject: Landscape Architecture
Topic: Low Carbon Landscape
Category: The Best Humanist Prize in Design Group of the 2013 "Yuan Ye Award" International Landscape Architecture Graduation Project/Thesis Competition

作品简介：

　　本项目运用可持续发展原理对农民工工地生活空间进行景观设计（图01-02），力求给农民工提供积极的生活方式，创建人与人之间亲近的交流平台，建立积极健康的生活氛围，满足农民工生活娱乐的需求，这样不仅可以在低成本、勤俭节约的前提下满足民工门的需求而且也促进了民工之间的交流活动（图03-06），提供一个经济实用性强、低碳、低技术的"高质量"生活（图07-11），改善农民工在工地生活空间的质量。

Introduction:

Based on the principle of sustainable development, this project designs the landscape for the living space of a rural migrant workers site (Fig.01- 02). The design represents an attempt to provide migrant workers with an active lifestyle by building a nearby communication platform at low cost and creating a positive and healthy living atmosphere, and thus meeting the living and entertainment requirements of migrant workers (Fig.03-06). The project aims to provide a "high-quality" life through economical, practical, low-carbon and low-technology means for the migrant workers living on the site(Fig.07-11).

竞赛作品 CONTEST ENTRIES

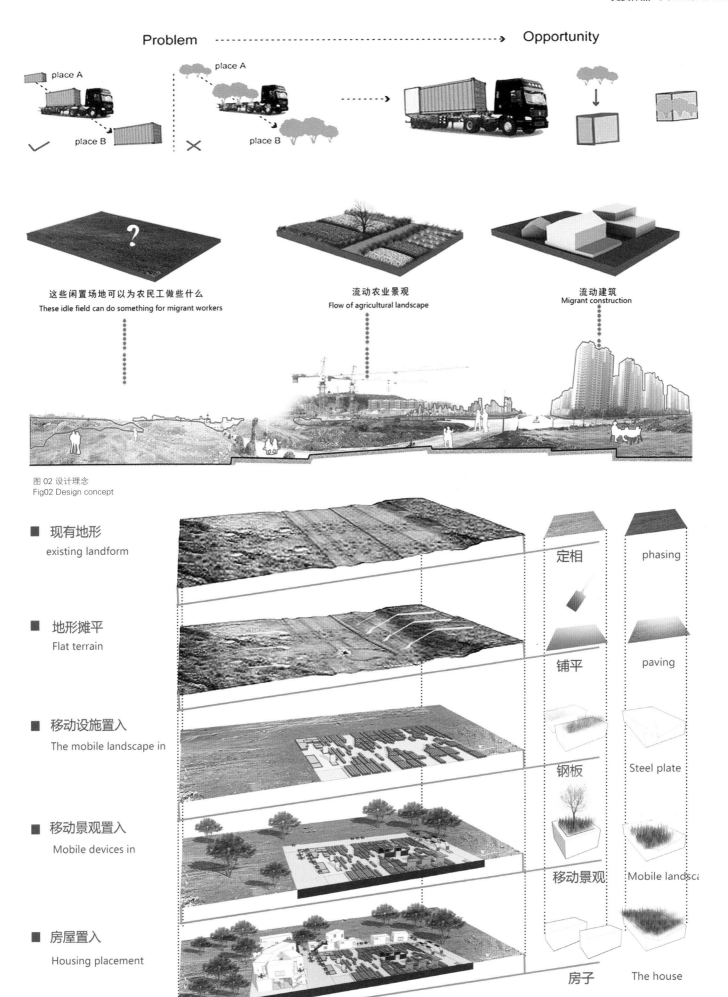

图 02 设计理念
Fig02 Design concept

图 03 竖向分析
Fig03 Vertical analysis

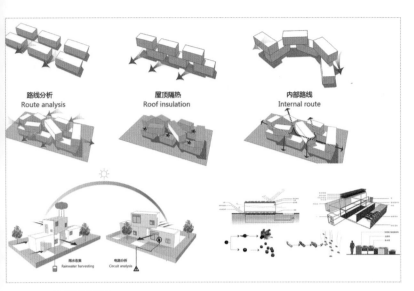

图04 建筑分析
Fig04 Architectural analysis

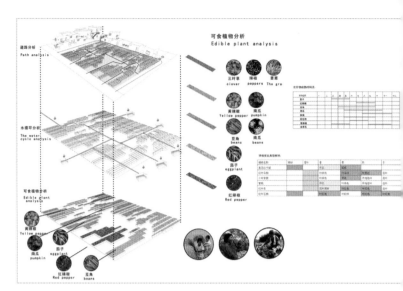

图05 植被分析
Fig05 Vegetation analysis

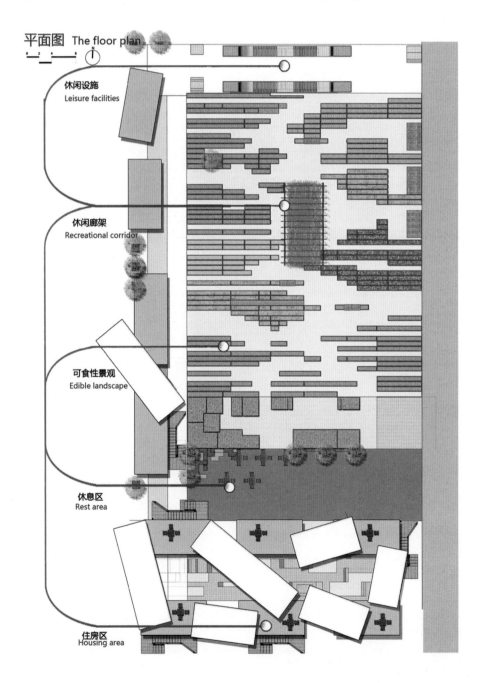

图06 平面图
Fig06 Floor plan

竞赛作品 CONTEST ENTRIES

图 07 居住区效果图
Fig07 Rendering of residential zone

图 08 廊架效果图
Fig08 Rendering of corridor

图 09 生活区效果图
Fig09 Rendering of living quarters

图 10 休息区效果图
Fig10 Rendering of rest area

图 11 鸟瞰图
Fig11 Bird's eye view

查尔斯·沙（校订）
English reviewed by Charles Sands

缘起天开
TianKai Story

"虽由人作，宛自天开"是中国古代造园大师计成在《园冶》中的名句，道出了造园的最高境界。以质量立身，以景效立命的天开集团恪守这份造园理念，抱着"为有限的城市空间营造无限的自然理想"的初衷，励志将公司打造成为"人居环境景观营造"的标杆式企业。天开园林首家公司由现任总裁陈友祥和副总裁谭勇于2003年在重庆联手创立，从此开启了天开时代的奋斗历程。2004年北京天开公司成立，天开集团进一步加快前进步伐，业务量逐年稳步增长，公司知名度也在大幅提升。2012年，天开完成了全国布局和业务升级，分公司覆盖中华各大区，北京、天津、上海、重庆、成都、长沙、哈尔滨等地。源于对景观工程的质量和景效的高要求，天开在行业内获得众多如万科、龙湖、纳帕、中旭、泰禾等高端合作伙伴的信赖，2012年3月，蓝光和骏地产也正式与天开签约结为战略联盟。时至今日，天开已发展成为一家集园林景观设计、园林工程施工、家庭园艺营造、苗木资源供应及石材加工为一体的领军式园林公司。我们的实力、品质和战略布局，能为客户有效降低沟通管理成本，更是全国地产企业所有项目卓越品质的保障。

China has a long history in Gardens and Landscape. "Materpiece of Nature, although Artificial Gardens", as the first and the most acknowledged philosophy about garden building in the world, was initiated from the book Yuanye which was written by Ancient Chinese gardening master Jicheng, and Tiankai was named after it. Tiankai company was established in Chongqing in 2003 by President Chen youxiang and Vice-president Tan yong. We come from nature and back to natue, that's the reason why we love nature gardens. Tiankai hopes to create infinite natural idea for the limited city. Today, Tiankai has become to comprehensive group corporation including garden construction, design, plants maintenance, seedling resources and so on. Tiankai has the second class qualification of garden construction, and B class qualification of landscape design. There are many international and outstanding designers in Tiankai. Tiankai is the best company in China in garden design and construction field. Tiankai has established 10 branches respectively in Beijing, Shanghai, Tianjin, Chongqing, Chengdu, Changsha and Haerbin. Tiankai's contruction business covers all over the country. The partners of Tiankai include Vanke, LongFor, Napa and Blue Ray, which are all very famous enterprises in China. We can effectively save communication and management cost for our clients and keep all the projects in consistent and excellent quality.

www.tkjg.com   tiankai@tkjg.com

三等奖 / THE THIRD PRIZE

行走的花园模块
——中关村二小校园景观设计
WALKING GARDEN MODULE
——ZHONGGUANCUN SECOND PRIMARY SCHOOL CAMPUS LANDSCAPE DESIGN

中文标题：行走的花园模块——中关村二小校园景观设计
组　　别：本科
作　　者：李　杨
指导教师：马晓燕　冯　丽
学　　校：北京农学院
学科专业：风景园林
研究方向：城乡景观规划设计
分　　类：2013"园冶杯"风景园林（毕业作品、论文）
　　　　　国际竞赛设计类作品三等奖

Title: Walking Garden Module——Zhongguancun Second Primary School Campus Landscape Design
Degree: Bachelor
Author: Yang Li
Instructor: Xiaoyan Ma, Li Feng
University: Beijing University of Agriculture
Specialized Subject: Landscape Architecture
Topic: Planning and Design of Urban and Rural Landscape
Category: The Third Prize in Design Group of the 2013 "Yuan Ye Award" International Landscape Architecture Graduation Project/Thesis Competition

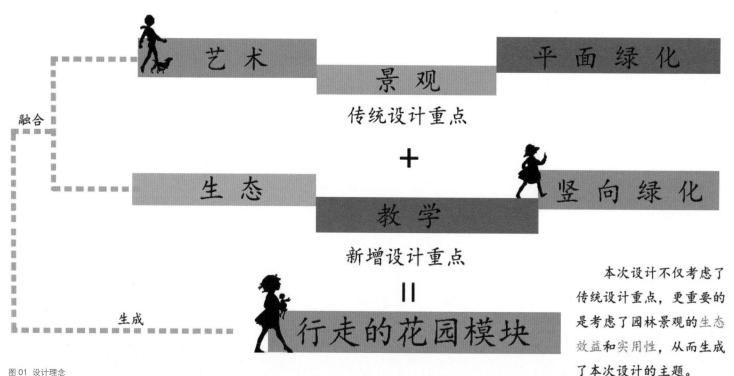

图01　设计理念
Fig01　The design concept

竞赛作品 CONTEST ENTRIES

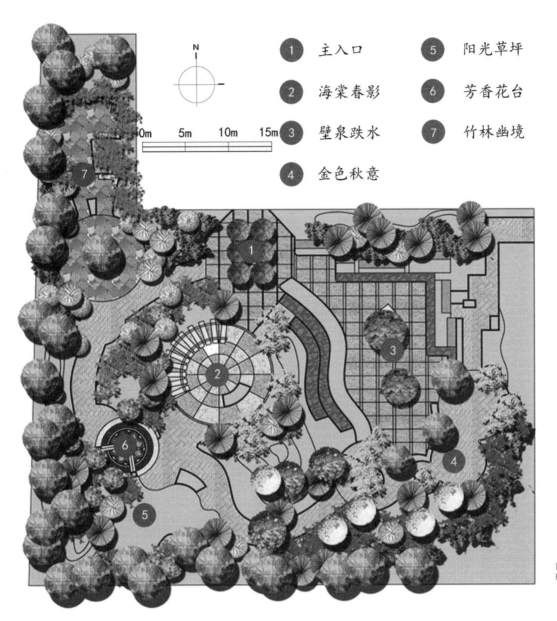

图 02 总平面图
Fig02 Master plan

作品简介：

校园是学生日常生活学习的重要场所，校园景观建设不仅要满足师生日常的活动需要，更重要的是要赋予校园景观以教育意义，使学生在其中学习知识，陶冶情操。因而，校园景观建设是值得人们深入研究和探讨的。

本次设计场地在北京市海淀区中关村二小百旺校区西南角的一块场地，紧挨学校操场和教职工宿舍，总面积为2587m²。设计不仅应用了传统景观美学的设计理念，更重要的是融入了生态性和实用性。该方案中将艺术性和生态性相结合、景观和教学结合、平面景观和竖向景观结合，这三个方面的结合充分体现了该园的设计理念（图01）。

该园由7个主要景观节点构成（图02）。虽然本次设计区域面积较小，但是能够在满足师生的休闲娱乐要求的同时，也能够发挥教学的功能，充分的体现了校园景观寓教于乐的功能。该园区主要的活动赏景区为海棠春影景观（图03-04）、壁泉跌水景观（图05-06）以及芳香花台景观（图07-08），除此以外，竹林幽境景观是该园林内的安静休息区（图09-10）。本次设计的重点景观是不同的水景，即壁泉、旱喷、镜面水池，分别从动态和静态两个方面表现了水体的景观美（图11-12），同时水景结合竖向空间设计，表现出了不同的空间氛围。

Introduction:

The campus is an important place for students in their daily lives. Landscape constructions on campuses should not only meet the daily needs of teachers and students, but should also play an educational and spiritual role. The site is on the southwest corner of Zhonguancun Primary School, Haidian. It is close to the school playground and faculty building, with a total area of 2587 square meters. The design combines functionality, artistry, ecology and education in both flat and vertical landscape elements. The combination of these three elements fully embodies the design concept of the park (Fig.01).

The park consists of seven main landscape nodes (Fig.02). Although the area is small, it can meet the leisure and entertainment needs of teachers and students, and also play an educational role. The main scenic spots for activities in the park are 'begonia spring shadow landscape' (Fig.03-04), 'wall fountain plunge landscape' (Fig.05-06) and 'fragrant flower-standing landscape' (Fig.07-08). In addition, the secluded 'bamboo forest landscape' provides a quiet rest area (Fig.09-10). The landscape is designed to focus on different waterscapes: wall fountain, dry spray and mirror pool, showing different aspects of the beauty of water both dynamic and static (Fig.11-12).

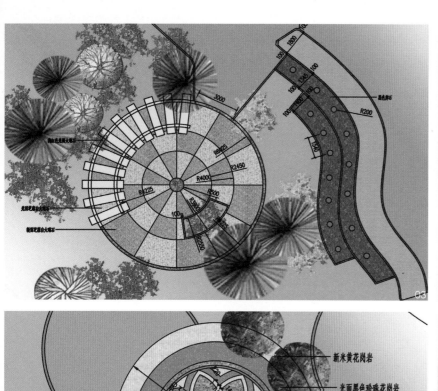

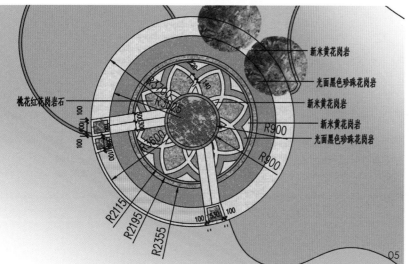

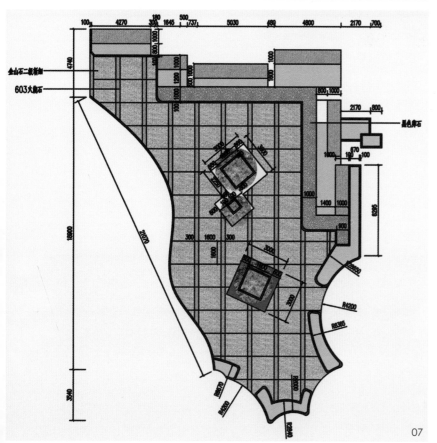

图 03 海棠春影景观平面图
Fig03 Begonia spring shadow landscape plan

图 04 海棠春影景观效果图
Fig04 Begonia spring shadow landscape rendering

图 05 壁泉跌水景观平面图
Fig05 Wall fountain plunge landscape plan

图 06 壁泉跌水景观效果图
Fig06 Wall fountain plunge landscape rendering

图 07 芳香花台景观平面图
Fig07 Fragrant flower-stand landscape plan

图 08 芳香花台景观效果图
Fig08 Fragrant flower-stand landscape rendering

图 09 竹林幽境景观平面图
Fig09 Secluded bamboo forest landscape plan

图 10 竹林幽境景观效果图
Fig10 Secluded bamboo forest landscape rendering

图 11 壁泉效果图
Fig11 Wall fountain rendering

图 12 镜面水池效果图
Fig12 Mirror pool rendering

查尔斯·沙（校订）
English reviewed by Charles Sands

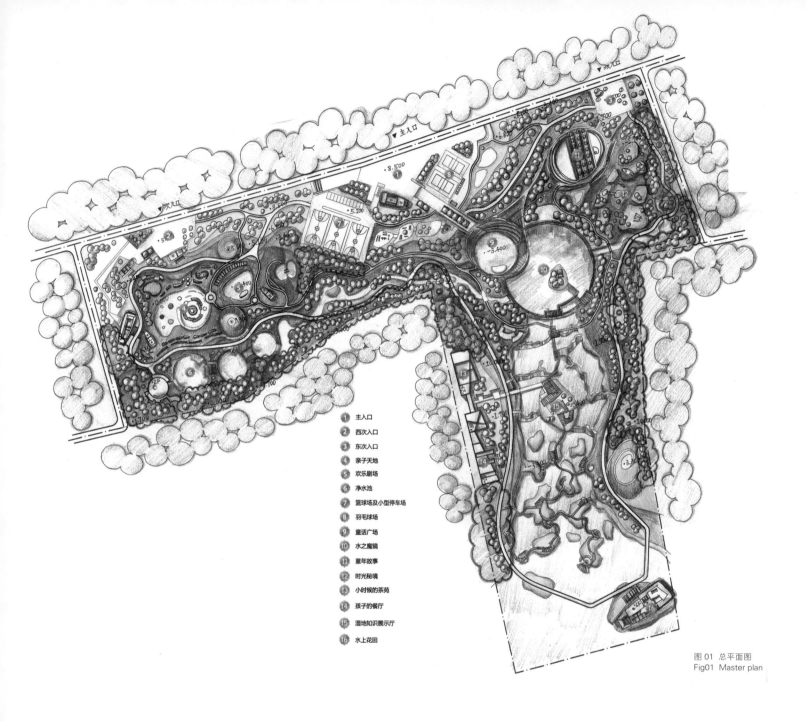

图01 总平面图
Fig01 Master plan

武汉市黄陂区船舶主题儿童公园设计
DESIGN OF THE SHIP THEMED CHILDREN'S PARK IN HUANGPI DISTRICT, WUHAN

中文标题：武汉市黄陂区船舶主题儿童公园设计
组　　别：本科
作　　者：刘 悦 林 黛（同为第一作者）
指导教师：戴 菲
学　　校：华中科技大学
学科专业：景观学
研究方向：景观设计
分　　类：2013"园冶杯"风景园林（毕业作品、论文）国际竞赛设计类作品三等奖

Title: Design of The Ship Themed Children's Park in Huangpi District, Wuhan
Degree: Bachelor
Author: Yue Liu, Dai Lin (Both are the first author)
Instructor: Fei Dai
University: Huazhong University of Science and Technology
Specialized Subject: Landscape
Topic: Landscape Design
Category: The Third Prize in Design Group of the 2013 "Yuan Ye Award" International Landscape Architecture Graduation Project/Thesis Competition

作品简介：

本设计充分考虑使用者的切实需求，结合周边船厂与地形地貌要素，创建了一座集功能、生态与设计特色于一体的儿童公园（图01）。设计利用场地中散落的废弃船只，根据需要改造成小品、栈桥等构筑物，凸显场地特色（图02-03），并利用理化与生物相结合的方式净化了船厂排出的污水（图04）。为柔化长江边的水泥堤坝，同时提高场地内外可达性，本方案通过挖填方式将原有的一马平川改造成结合堤坝层层下落的台地。最后根据周边使用者年龄层次分布，对公园进行分区（图05-07），有针对性地设计节点（图08-10）。

Introduction:

The Children's park is designed by considering the real needs of the users, and combining elements of the surrounding shipyards and local topography. Functionality and ecology are integrated into the design features (Fig.01). The abandoned ships scattered around the place are recycled and transformed into scenic bridges and other elements, which highlight the characteristics of the place (Fig.02-03). Moreover, physical and chemical processes are coordinated with biological processes to purity the polluted water from the shipyard (Fig.04). The project also enlivens the flat topography with a terraced garden that backs onto the dam. Through cut and fill the edge of the concrete structure is softened and accessibly is improved. Finally, the park is divided into different areas based on the different ages of the users (Fig.05-07) with landscape nodes designed accordingly (Fig.08-10).

图02 废弃船只改造一
Fig02 Transformation of abandoned ship one

图03 废弃船只改造二
Fig03 Transformation of abandoned ship two

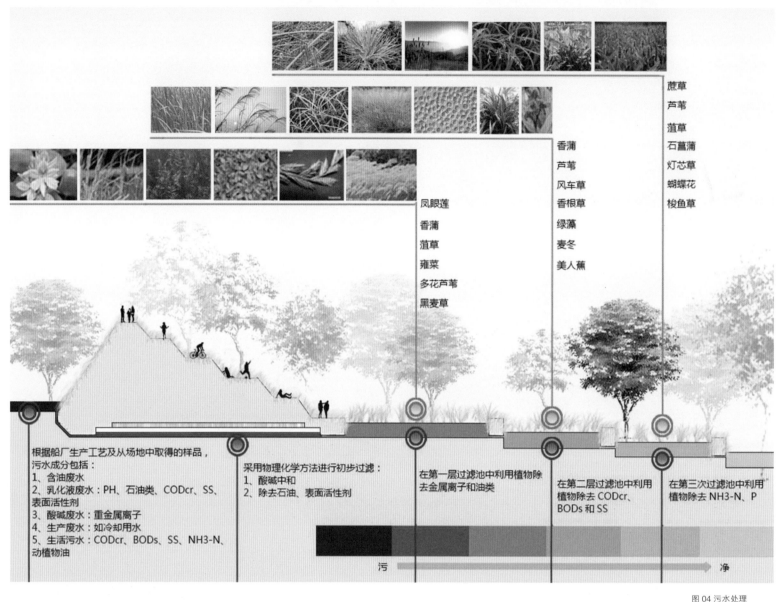

图04 污水处理
Fig04 Sewage disposal

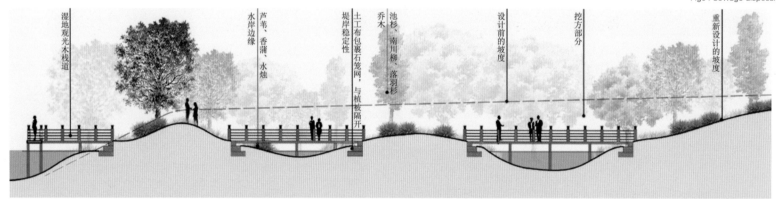

图05 挖方示意
Fig05 Cut

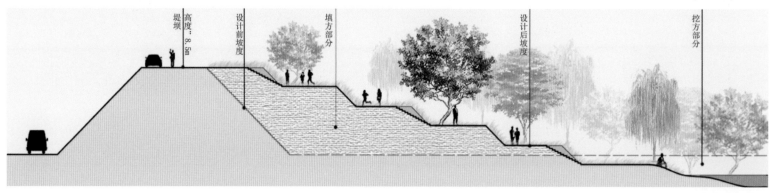

图06 填方示意
Fig06 Fill

竞赛作品 **CONTEST ENTRIES**

图08 活力运动场效果
Fig08 Exciting area

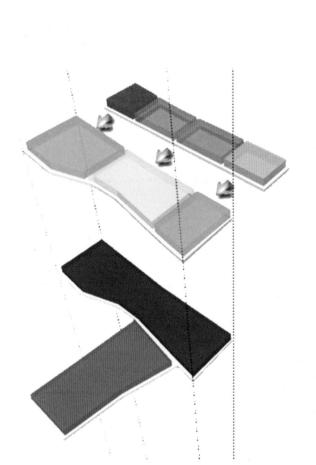

■ 静态活动区
■ 动态活动区
■ 幼儿活动区
■ 青少年运动区
■ 儿童游乐区
■ 幼儿园
■ 中学
■ 小学

图07 功能分区
Fig07 Functional division

图09 儿童游乐园效果
Fig09 Children's playground

图10 污水处理区效果
Fig10 Sewage disposal area

查尔斯·沙（校订）
English reviewed by Charles Sands

曲水花洲

——南宁青秀山水生花园概念设计

BENDING WATER AND FLORAL SANDBAR

——NANNING · QINGXIU MOUTAIN HYDROPHYTE GARDEN CONCEPTUAL DESIGN

中文标题：曲水花洲——南宁青秀山水生花园概念设计
组　　别：本科
作　　者：孙林琳
指导教师：沈守云
学　　校：中南林业科技大学
学科专业：园林
研究方向：规划设计
分　　类：2013"园冶杯"风景园林（毕业作品、论文）
　　　　　国际竞赛设计类作品三等奖

Title: Bending Water and Floral Sandbar——Nanning · Qingxiu Moutain Hydrophyte Garden Conceptual Design
Degree: Bachelor
Author: Linlin Sun
Instructor: Shouyun Shen
University: Central South University of Forestry and Technology
Specialized Subject: Landscape
Topic: Landscape Planning and Design
Category: The Third Prize in Design Group of the 2013 "Yuan Ye Award" International Landscape Architecture Graduation Project/Thesis Competition

图01 青秀山现状手绘图（底图来源：南宁市规划管理局）
Fig01 Present situation map (Source: Nanning Planning Bureau)

竞赛作品 CONTEST ENTRIES

■ 图例

1. 主入口
2. 游船码头
3. 人造树屋
4. 架空走廊
5. 滩涂沙地
6. 凤仙花花田
7. 毛茛科花田
8. 一二年生花径
9. 郁金香广场
10. 次入口
11. 湿地体验区
12. 雨久花科花田
13. 观景亭
14. 叠水瀑布
15. 次入口
16. 荷田栈道
17. 花艺展馆
18. 展馆入口

图 02 总平面图
Fig02 Master plan

作品简介：

本案基址位于广西南宁青秀山森林植物园内，利用当地的地理、生态优势，建立水生湿地花园（图 01-02）。该设计以"曲水花洲"为主题，以水体设计和植物配置为设计重点。通过架空走廊和游船的路线设计，为游客提供高空俯视、亲临水中的多重视角、多重感官体验（图 03-05）。同时，以"入梦——溺梦——梦醒——寻梦"这一故事性游线引导游客完成全园的游览参观（图 06-08）。独特的生态体验区和花艺展馆，让游客在观赏游憩的同时，得到健身保健和科普教育的体验（图 09-11）。

Introduction:

The site is located in the forest Botanical Garden at Qingxiu Mountain, Nanning, Guangxi. Taking advantage of the local geography and ecology, an aquatic wetland garden is established(Fig.01-02). The design takes 'bending water and floral sandbar' as its theme, focusing on water body design and plant distributions. The elevated corridors and sightseeing boat with a variety of routes offer visitors a variety of viewing angles(Fig.03-05). Tourists are led through a series of experiences termed "arriving in the dream - indulging in the dream – waking up – seeking dreams". Thus tourists are led through narrative theme in their experience of the garden (Fig.06-08). This series of ecological experiences as well as the floral hall provide tourists with an invigorating and educational experience(Fig.09-11).

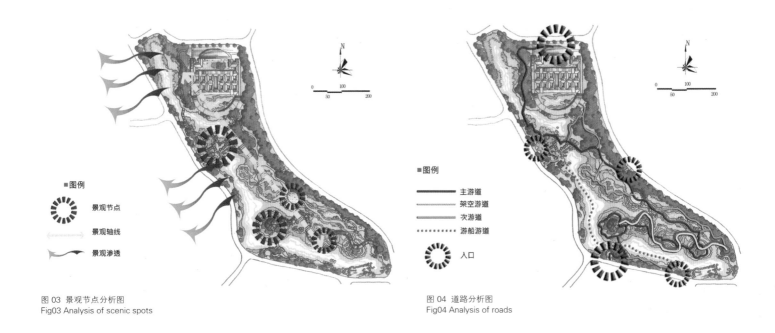

图 03 景观节点分析图
Fig03 Analysis of scenic spots

图 04 道路分析图
Fig04 Analysis of roads

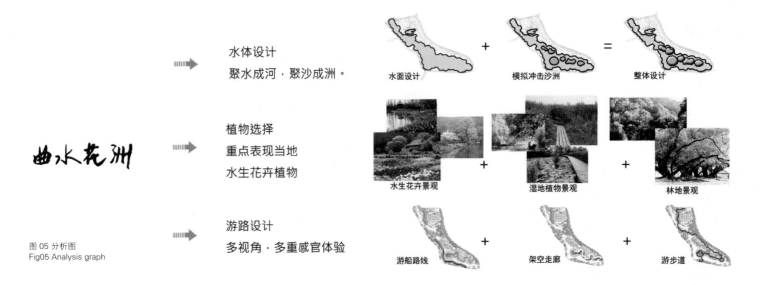

图 05 分析图
Fig05 Analysis graph

图 06 断面图
Fig06 Cross-section graph

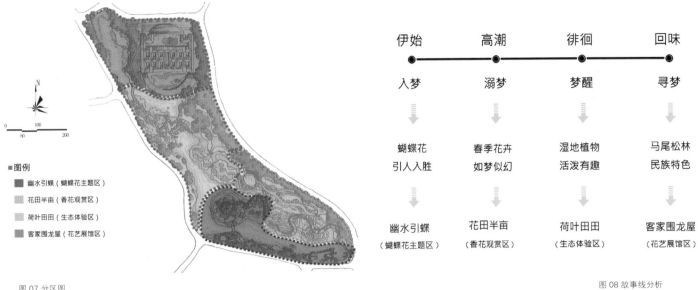

图 07 分区图
Fig07 Area analysis

图 08 故事线分析
Fig08 Story line analysis

图 09 幽水引蝶
Fig09 Butterflies attracted by the stream

图 10 花田半亩
Fig10 The flower field

图 11 生态体验区
Fig11 The ecological experience area

查尔斯·沙（校订）
English reviewed by Charles Sands

2012中国优秀园林工程金奖——中国铁建·国际城（杭州）示范区园林绿化工程

杭州市园林绿化工程有限公司
HangZhou Landscape Garden Engineering Co.,Ltd.

企业资质

城市园林、市政公用工程施工总承包壹级	
园林古建筑工程专业承包贰级	城市及道路照明工程专业承包壹级
风景园林工程设计专项乙级	建筑装饰工程设计专项乙级
绿化造林设计施工乙级	房屋建筑工程施工总承包叁级
土石方工程专业承包叁级	建筑装修装饰工程专业承包叁级

大入部合源律入居佳境　长青基业缔造品质生活

杭州市园林绿化工程有限公司创建于1992年，现已发展成集园林市政规划设计、工程项目施工、花卉种苗研发生产为一体的企业集团，拥有近十家全资子公司和控股股公司，业务范围遍及全国。

杭州园林以创新进取的态势拓展多元绿色产业，加快市场拓展步伐，逐步成长为行业的领军企业，位居建设部白皮书大型项目施工能力排名全国第二名，综合排名全国十强企业；具备国家园林一级、市政工程施工总承包一级、风景园林设计专项乙级等十余项资质。

通过ISO9001质量管理、ISO14001环境管理、OHSAS18001职业健康安全管理体系认证。工程施工年产值超十亿元，承建的工程先后获得"鲁班奖"（国家优质工程）、"中国风景园林金奖"、"钱江杯"、"省、市优秀园林绿化工程金奖"等众多荣誉。先后被评为"国家高新技术企业"、"全国城市园林绿化企业"、"全国农ün系统劳动关系和谐企业"、"中国十大创新型风景园林观赏苗木企业"、"省级林业重点50强"、"AAA级守合同、重信用"龙头企业"、"等荣誉。连续12年荣膺浙江省"AAA级守合同、重信用"企业。

杭州园林为浙江省林木种质资源保育与利用公共基础条件平台重要成员、种苗研发基地被授予"中国桂花品种繁育中心"、"杭州市高新技术木种质研发中心"等荣誉称号，组织攻夫的杭州市政府重点科技项目《桂花品种基因库建设及优良品种选育推广》荣获2007年度"杭州市科技进步一等奖"；2011年3月，公司实施并完成了浙江省重大科技项目《桂花种质创新及促成栽培关键研究与示范》。

企业承担了大量的社会责任，履行着十余家行业协会的领导职能，如中国花协桂花分会、中国花协绿化观赏苗木分会、浙江省林学会、浙江省植物学会园林植物分会、浙江省风景园林学会、浙江省花协绿化苗木分会等。

公司地址：杭州市凯旋路226号浙江省林业厅6F/8F　联系电话：0571-86095666/86431126　传真：0571-86097350　邮编：310020　网站：www.hzylh.com

盛世绿源
彩色城市

为什么说盛世绿源是专业的彩色树供应商？

1、种苗自北美原生区引进，片植、孤植均有极好的造景效果

2、五大彩色树种植基地：总圃面积2万余亩，苗木50余万株，满足客户的需求

3、不同温度带的适生基地，保障了树木在国内各地区能够健康生长

4、完备的技术措施，保证了苗木的成活率

红枫、红栎、海棠、白蜡、山楂

诚信 责任 卓越

联系我们：0411-83705888
　　　　　　　83670830
　　　　　　　83659060

盛世绿源科技有限公司
大连市西岗区北京街126号

鼓励奖 / THE HORORABLE MENTION

马头山国家级自然保护区生态旅游区规划设计
PLANNING OF MATOUSHAN NATIONAL NATURE RESERVE ECOTOURISM ZONE

中文标题：马头山国家级自然保护区生态旅游区规划设计
组　　别：本科
作　　者：杨天人　常　凡　李　菁
指导教师：吴承照　陈　静
学　　校：同济大学
学科专业：景观学系
研究方向：城乡人居环境可持续发展、景观游憩学
分　　类：2013"园冶杯"风景园林（毕业作品、论文）
　　　　　国际竞赛规划类作品鼓励奖

Title: Planning of Matoushan National Nature Reserve Ecotourism Zone
Degree: Bachelor
Author: Tianren Yang, Fan Chang, Jing Li
Instructor: Chengzhao Wu, Jing Chen
University: Tongji University
Specialized Subject: Landscape
Topic: Sustainable Development of Urban and Rural Human Settlement, Landscape
Category: The Honorable Mention in Planning Group of the 2013 "Yuan Ye Award" International Landscape Architecture Graduation Project/Thesis Competition

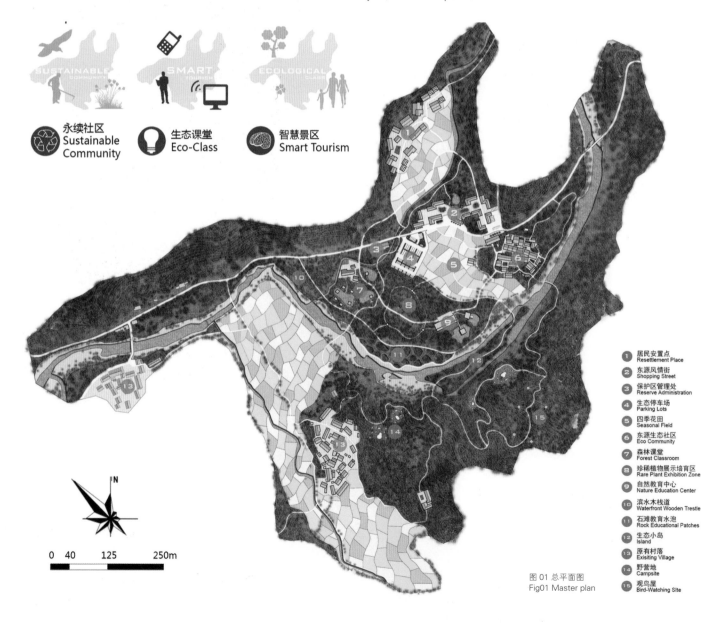

图01 总平面图
Fig01 Master plan

1. 居民安置点 Resettlement Place
2. 东源风情街 Shopping Street
3. 保护区管理处 Reserve Administration
4. 生态停车场 Parking Lots
5. 四季花田 Seasonal Field
6. 东源生态社区 Eco Community
7. 森林课堂 Forest Classroom
8. 珍稀植物展示培育区 Rare Plant Exhibition Zone
9. 自然教育中心 Nature Education Center
10. 滨水木栈道 Waterfront Wooden Trestle
11. 石滩教育水泡 Rock Educational Patches
12. 生态小岛 Island
13. 原有村落 Exisiting Village
14. 野营地 Campsite
15. 观鸟屋 Bird-Watching Site

功能分区图
Zoning

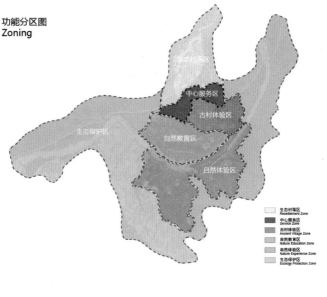

交通分析图
Transportation Analysis

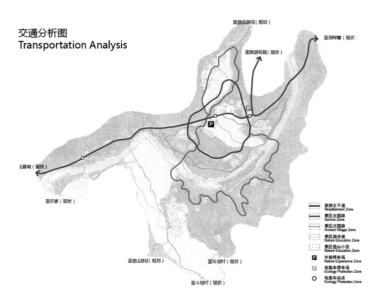

教育游径规划
Eduacational Trail

图 02 规划分析图
Fig02 Planning analysis

种植规划
Planting Design

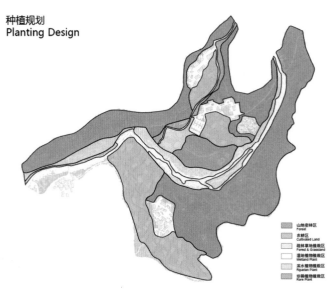

作品简介：

江西马头山自然保护区位于武夷山脉中段腹地，地处资溪县北部与鹰潭相邻，北与龙虎山相连，南靠大觉山。其中生态旅游区范围跨越保护区边界，包含了部分实验区及保护区外围地带，生态环境优良，生物多样性丰富。本文对自然保护区实验区如何协调保护与发展的矛盾，以及自然保护区外围地带如何利用保护区知名度发展相适应的经济产业，带动保护区发展做出探讨性研究，为我国自然保护区实验区及保护区外围地带生态旅游的发展提供了新的思路，并将理念应用到马头山自然保护区生态旅游区的总体规划中（图01）。在总体规划部分完成了现状条件分析、规划原则与依据、规划结构与布局、道路交通规划、主题游线规划、居民社会调控规划、土地利用协调规划等（图02）。在详细规划设计层面，选择位于保护区边界外围及保护区实验区交界处的东源景区，从功能分区、交通系统、项目策划等层面进行分析与规划设计，并对其中的若干节点进行重点设计（图03-09）。

Introduction:

Matoushan National Nature Reserve is located in Wuyi Mount Range, Zixi County, Jiangxi Province, north to Mount Longhu and south to Mount Dajue. The ecotourism zone is across the boundary of the nature reserve, including part of the experimental area and the periphery of the nature reserve, which has a beautiful ecological environment and abundant biodiversity. This article will research how to coordinate the contradiction between protection and development in experimental areas and how to take advantage of the popularity of nature reserves to promote the local economy. It will bring some new ideas to the development of national nature reserve ecotourism in our country, which will be also used in the master planning of Matoushan National Nature Reserve (Fig.01). For the master planning, we completed an analysis of the present situation, planning principles and references, planning structure and zoning, transportation system planning, thematic touring route planning, the residential planning, land use planning and so on (Fig.02). In the detailed planning and design field, we selected Dongyuan scenic area to analyze and plan the zoning, transportation system design, tourism program planning etc., which is on the periphery of the reserve close to the experimental area. As well much attention was payed to designing nodes in this area(Fig.03-09).

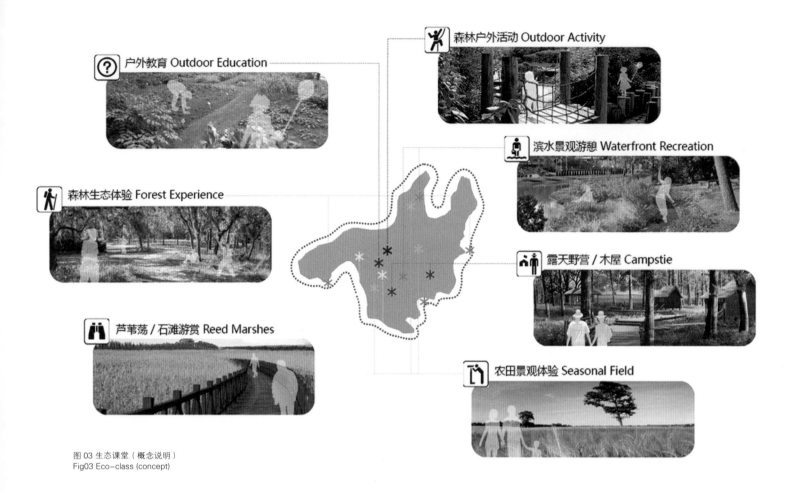

图 03 生态课堂（概念说明）
Fig03 Eco-class (concept)

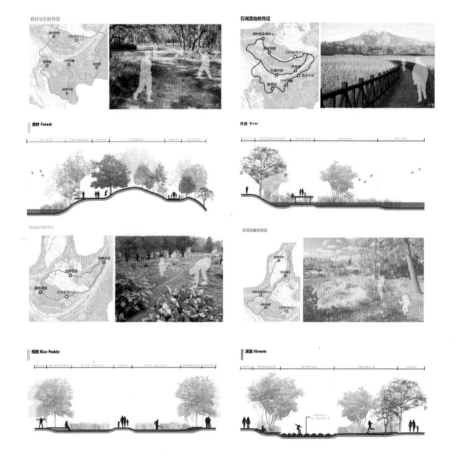

图 04 生态课堂（剖面图）
Fig04 Eco-class (section)

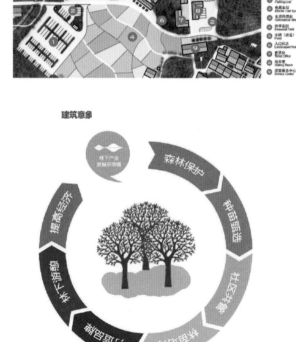

图 05 永续社区—商—户模式
Fig05 Sustainable community-commercial mode

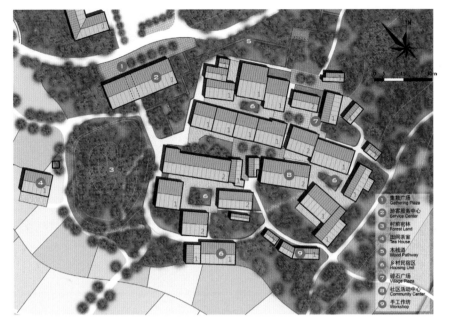

图 06 永续社区（总平面图）
Fig06 Sustainable community (master plan)

图 07 智慧景区（概念说明）
Fig07 Scenic wisdom (concept)

图 08 永续社区（分析图）
Fig08 Sustainable community-home stay (analysis)

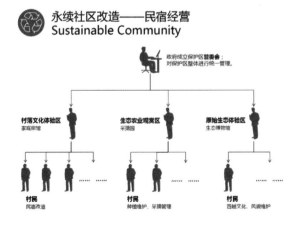

图 09 永续社区民宿经营（概念说明）
Fig09 Sustainable community-home stay (concept)

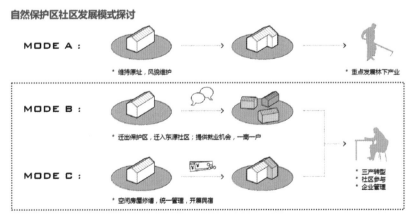

查尔斯·沙（校订）
English reviewed by Charles Sands

田园城市
——北京上地树村有机更新设计
ORGANIC GARDEN CITY
——BEIJING SHANGDI TREE VILLAGE RENEWAL DESIGN

中文标题：田园城市——北京上地树村有机更新设计	Title: Organic Garden City——Beijing Shangdi Tree Village Renewal Design
组　　别：本科	Degree: Bachelor
作　　者：王艳秋　武　键	Author: Yanqiu wang, Jian Wu
指导教师：李　婧　姬凌云	Instructor: Jing Li, Lingyun Ji
学　　校：北方工业大学	University: North China University of Technology
学科专业：城乡规划	Specialized Subject: Urban and Rural Planning
研究方向：城市设计	Topic: Urban Planning
分　　类：2013"园冶杯"风景园林（毕业作品、论文）国际竞赛规划类作品鼓励奖	Category: The Honorable Mention in Planning Group of the 2013 "Yuan Ye Award" International Landscape Architecture Graduation Project/Thesis Competition

图 01 鸟瞰图
Fig01 Bird's eye view

竞赛作品 CONTEST ENTRIES

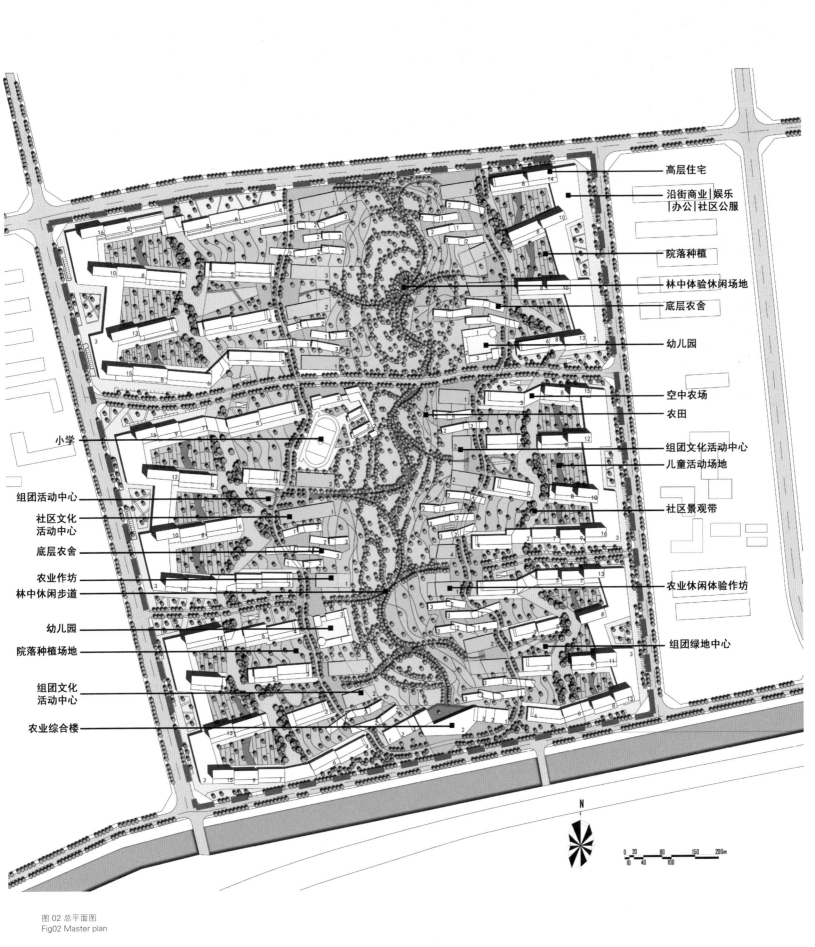

图 02 总平面图
Fig02 Master plan

图03 产业结构图
Fig03 Industrial structure

图04 组织体系图
Fig04 Organizational system

图05 生活方式图
Fig05 Mode of life

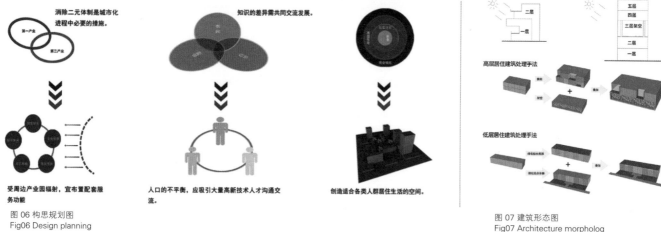

图06 构思规划图
Fig06 Design planning

图07 建筑形态图
Fig07 Architecture morpholog

作品简介：

本方案根据现代城市化发展速度加快，"城中村"服务设施、居住环境、产业模式等一系列与城市发展不协调问题进行设计，欲创造一个集"工、农、创"为一体的多元化综合社区。通过综合社区的建设，将解决绿色生态与城市环境相矛盾的空间问题（图01-02），同时协调不同人群发展与城市发展的诸多社会问题（图03-06）。

田园中的城市是将城市中的建筑、商业、服务、文化等多方面的元素规划在以生态、绿色、环保为主题的田园社区中（图07）。

城市中的田园社区是将农田、农舍、农场等一系列田园的生活设施与城市中的建筑相融合（图08-11）。

Introduction:

According to the rapid speed of modern urbanization, this plan is designed based on a series of problems endemic to urban development, including service facilities, living environments, and industrial pattern in "villages inside cities". The goal is to create a diversified comprehensive community, which means an integrated set of "industry, agriculture, and creation". Through comprehensive community construction, the spatial problem of contradiction between green ecology and the urban environment will be mitigated (Fig.01-02), and the social problem of coordinating human development and urban development will be mitigated as well (Fig.03-06).

The idea is to create a city in a garden by planning the urban construction, commerce, services, culture and other various elements in the rural communities to be themed by ecology, urban greening, and environmental protection (Fig.07).

Rural communitues inside the city are intergrated into a series of rural living facilities, which include farmhouses and agricultural land, becoming an integral part of the surrounding urban construction (Fig.08-11).

查尔斯·沙（校订）
English reviewed by Charles Sands

竞赛作品 CONTEST ENTRIES

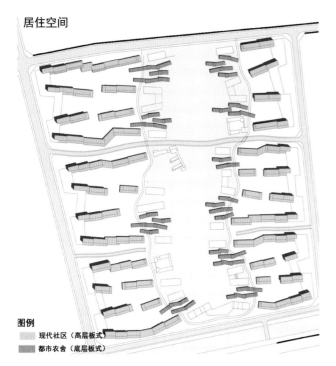

图 08 居住空间分析图
Fig08 Analysis of living space

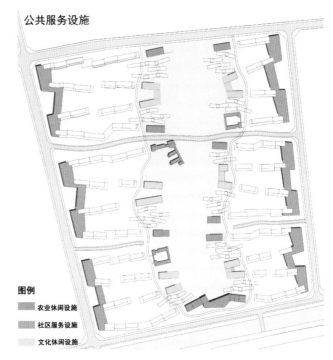

图 09 公共服务设施分析图
Fig09 Analysis of public service facilities

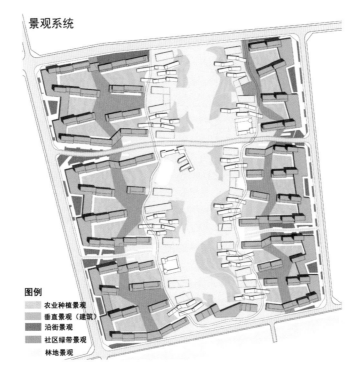

图 10 景观系统分析图
Fig10 Analysis of landscape system

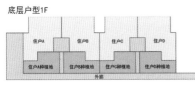

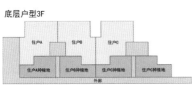

图 11 建筑平面图
Fig11 Floor plan

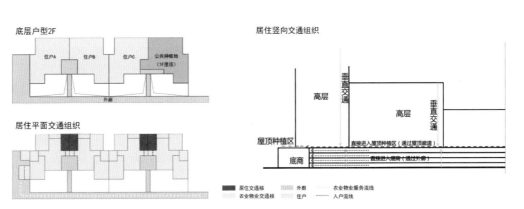

WORLDSCAPE No.1 2014   159

图 01 总体布局
Fig01 The overall layout

耕·记—城市绿色加工厂
——重庆渝北区生态苗木产业园

URBAN GREEN PLANTS
——ECOLOGICAL INDUSTRIAL PARK OF SEEDLINGS IN CHONGQING YUBEI DISTRICT

中文标题：耕·记——城市绿色加工厂——重庆渝北区生态苗木产业园
组　别：本科
作　者：余 梅　李今朝　李月文
指导教师：张建林
学　校：西南大学
学科专业：城市规划
研究方向：公园规划设计
分　类：2013"园冶杯"风景园林（毕业作品、论文）国际竞赛规划类作品鼓励奖

Title: Urban Green Plants—Ecological Industrial Park of Seedlings in Chongqing Yubei District
Degree: Bachelor
Author: Mei Yu, Jinzhao Li, Yuewen Li.
Instructor: Jianlin Zhang
University: Southwestern University
Specialized Subject: Unban Planning
Topic: Park Planning and Design
Category: The Honorable Mention in Planning Group of the 2013 "Yuan Ye Award" International Landscape Architecture Graduation Project/Thesis Competition

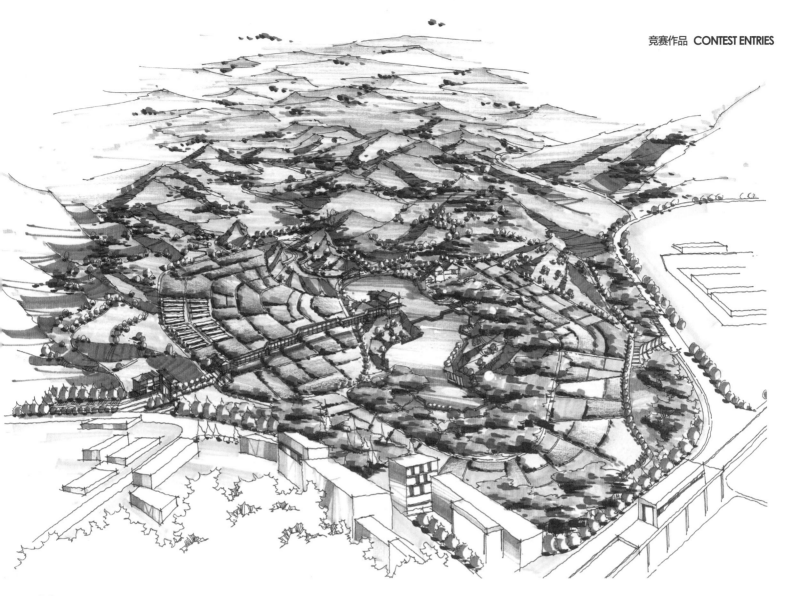

图 02 鸟瞰
Fig02 Bird's eye view

作品简介：

随着城市绿化建设的高速推进，苗木产业已逐渐成为城市近郊发展的一个新的增长点。都市苗木产业园兼顾苗木生产与观光游览两大特征。依托自然地形地貌，集苗木生产功能、园区游览功能、植物科普教育功能于一体。

基于以上发展需求，同时满足城市发展和近郊苗木产业的调整，规划选址于重庆市渝北区金开大道北侧，占地面积约33公顷的城郊绿地，并进行一系列的场地及场地周边分析（图01-04）。在建立以苗木生产为主，休闲游览为辅的生态苗木产业园的基础上提出"动态产业景观"这一概念，关注一个苗木生产周期内从"苗木栽植"到"苗木移植"的运作过程，通过由微观层面转变至宏观层面，着眼于整个苗木产业园在一个生产周期内的苗木更替循环过程。同时与城市绿化生态修复工程相互结合，将该产业园比作"城市绿色加工厂"，实现土地利用价值最大化，满足基本苗木生产的同时依照季相变化创造更具生命力的"动态更替景观"。并在其中配有供人游憩的景观节点，以及富有特色的景观建筑设施，以便充分挖掘生产性绿地的景观价值（图05-10）。

Introduction:

With the rapid growth of urban areas, the nursery stock industry has gradually become a new growth point of suburban development. Urban industrial parks provide opportunities for both seedling production and sightseeing. Following the natural topography, nursery stock production, park tours, and plant science education are combined together in one entity, thereby fully exploiting the landscape value of productive greenland.

Our plan is intended to meet development requirements, while allowing for growth in urban development and an expansion of the suburban nursery stock industry,The planning site is on the north side of JinKai road in Chongqing's Yubei District, covering an area of approximately 33 hectares of suburban Greenland (Fig.01-04). Concerned about the operation process of the seedling production cycle from planting seedlings to seedling transplants, the concept of "Dynamic Industry Landscape" is proposed.This involves building an ecological industry park of nursery stock seedlings, and supplementing this industry with recreational tours.

By transforming the micro level into the macro level, we pay attention to the replacement process of the nursery production cycle in the entire nursery industry park. Meanwhile combining ecological restoration with urban greening, maximizes the value of the land and creates a 'dynamic replacement landscape', accomplishing basic seedling production in accordance with seasonal change (Fig.05-10).

主题：花木环山、阡陌交通、渔人唱晚

（一）"花木环山"——设计以"山"为依托，强调由低至高层层叠加的肌理，运用于苗木生产用地的划分，顺应山体整体等高线走势，由下至上分层规划各类树种，打造丰富的山地立体景观。

（二）"阡陌交通"——取自陶潜《桃花源记》，意指农田地里的小道和灌溉渠道，是基地最为显著的要素。故设计将其串联各个苗木区，同时满足苗木生产灌溉需求。

（三）"渔人唱晚"——设计将"渔乡"作为中心游览区的文化元素，以"水"为脉络，以"鱼"为线索，沿湖打造集观鱼、钓鱼、戏鱼、听鱼、寻鱼、品鱼为一体的感观体验。

图 03 规划理念
Fig03 Planning ideas

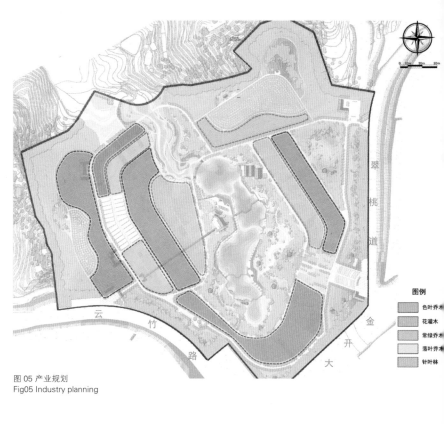

图 05 产业规划
Fig05 Industry planning

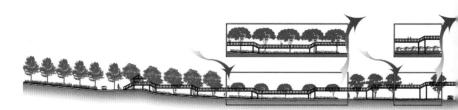

图 06 第一年种植图
Fig06 The plantingchart of the first year

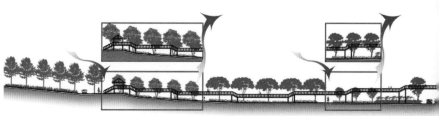

图 07 第二年种植图
Fig07 The plantingchart of the second year

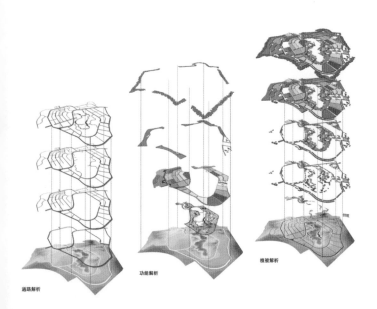

图 04 方案解析
Fig04 Program analysis

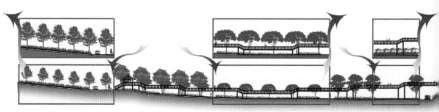

图 08 第三年种植图
Fig08 The plantingchart of the third year

图 09 第四年种植图
Fig09 The plantingchart of the fourth year

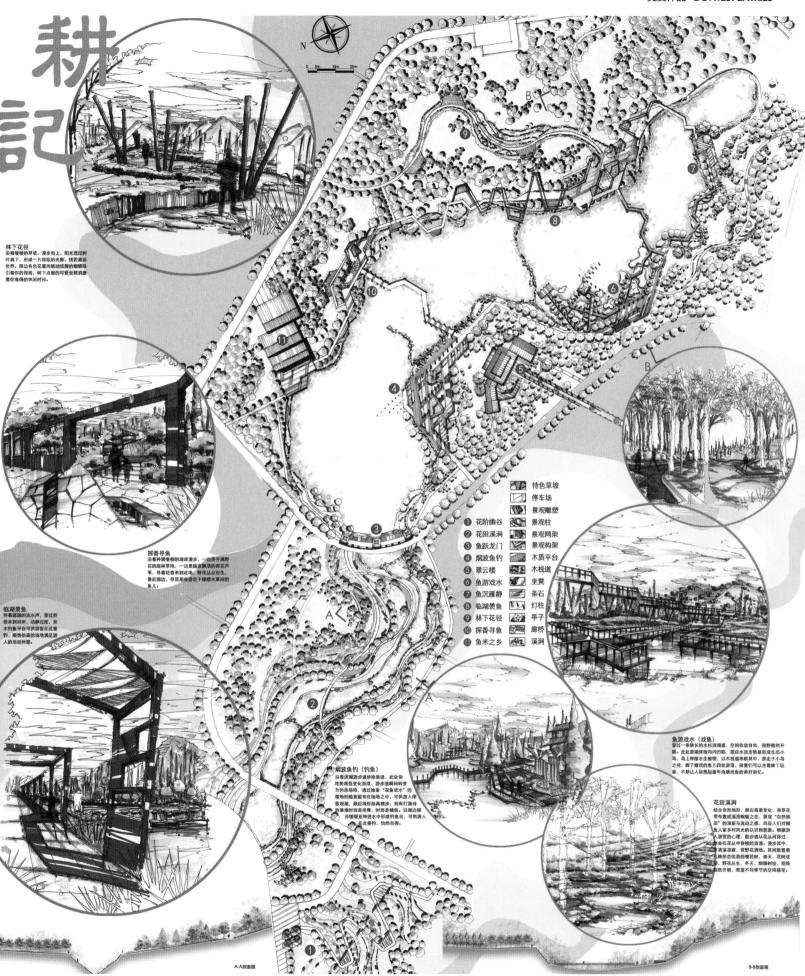

图10 景观节点总图
Fig10 Total landscape node map

LIGHT
——城市公共空间小型构筑体概念设计
LIGHT
——CONCEPTUAL DESIGN OF SMALL STRUCTURES IN URBAN PUBLIC SPACES

中文标题：LIGHT——城市公共空间小型构筑体概念设计
组　　别：本科
作　　者：曾舜怡
指导教师：吴宝娜
学　　校：华南农业大学
学科专业：城市规划
研究方向：城市公共空间优化
分　　类：2013"园冶杯"风景园林（毕业作品、论文）
　　　　　国际竞赛设计类作品鼓励奖

Title: LIGHT— Conceptual Design of Small Structures in Urban Public Spaces
Degree: Bachelor
Author: Shunyi Zeng
Instructor: Baona Wu
University: South China Agricultural University
Specialized Subject: Urban Planning
Topic: Optimization for Urban Public Space
Category: The Honorable Mention in Design Group of the 2013 "Yuan Ye Award" International Landscape Architecture Graduation Project/Thesis Competition

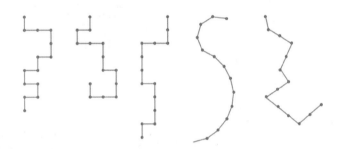

Stept 1:

直线 —打断→ 线段 —加厚拉高→ 平面 —旋转→ 空间

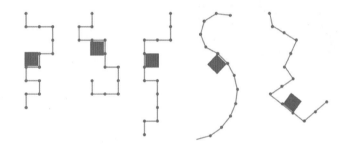

Stept 2:
为了增加其实用性和主题性，在围合起来的空间内添加构筑体块

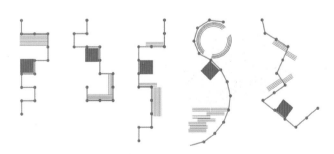

Stept 3:
不同的旋转方式衍生出大小形状各有不同的空间形态，
针对各种空间的特征给它们定义不一样的使用功能

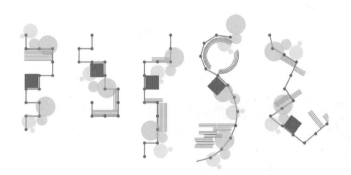

Stept 4:
可在休憩空间内适当增添种植，
但更推荐根据地块植物分布现状来分隔相适应的空间

图 01 空间划分的原理
Fig01 Principles for setting apart spaces

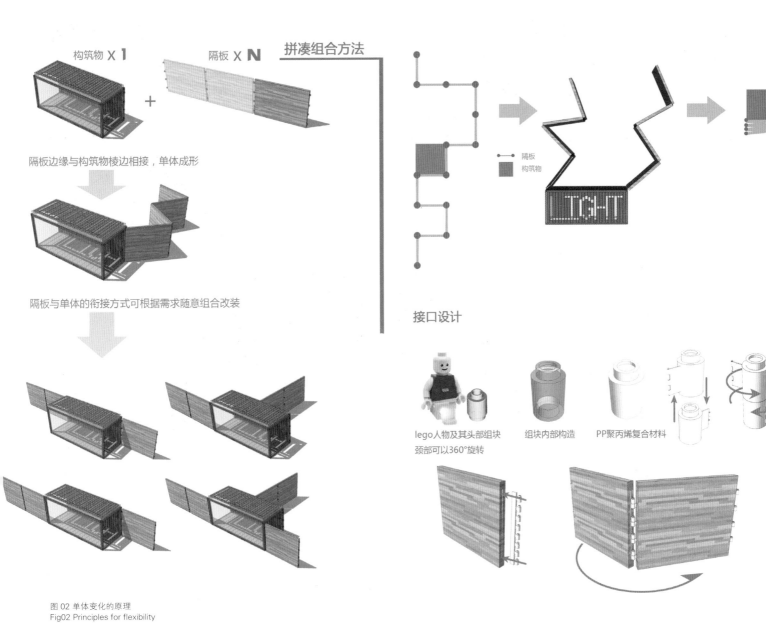

图02 单体变化的原理
Fig02 Principles for flexibility

作品简介：

本设计作品"LIGHT"是一个在不改变原用地现状的情况下，通过自身发生的魔术变化从而优化空间格局的小型构筑体，尤其针对于解决高密度城市里公共空间零碎和不平均的问题现状（图01-02）。设计由若干块隔板组成，构筑被固定放置后，与其连接的隔板可以根据用地的地形、大小、功能以及服务对象等需求进行旋转和拼接，从而围合成为多个相对独立的活动空，成为真正为人所设、供人所用的公共场所（图03-08）。设计其采用轻质复合型材料制作，符合未来的环保节能城市建设目标。

Introduction:

The project, "LIGHT" is a small structure which can change itself magically to optimize spatial patterns without disrupting the current state of land use. The project specifically aims to give site-specific solutions for fragmented and irregular public spaces (Fig.01-02).

The structure is constructed with partition boards. After it is fixed in a predetermined location, the linked clapboards can be rotated or spliced according to the terrain, scale, function and the demands of a given site. Thus, several relatively independent spaces are framed by the surrounding boards. These public spaces are designed for flexibility of space and function. They are intended to play an important role in the public environment (Fig.03-08). Being made of lightweight composite materials, the structures conform to the goal of constructing an environment-friendly and energy-saving society.

图 03 贴心的新设施
Fig03 Thoughtful facilities

图 04 人与环境的互动
Fig04 Interactions between people and the environment

线型模板

隔板的排列成直线型，并与构筑物平行于同一侧。对人群的活动有比较明显的限制，对隔板前后两边的环境有较好的隔离效果。

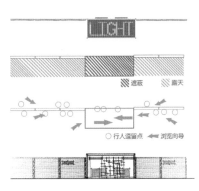

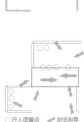
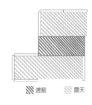

围合型模板

围合形模板的空间感较强；2~3快隔板便能围合成一个半开放私密空间。由于存在空间差异，因此这种模板允许不同空间有不同的景观特色，同理不同空间内所赋予的人流活动也可以不一样。

图 05 线型模型
Fig05 Linear type model

图 06 围合型模型
Fig06 Enclosure type model

遮蔽型模板

在围合型的基础之上，根据估计不同空间里的平均人流量和人群逗留时间长短，对部分重点位置添加晴雨顶棚。里面设计成为上也能根据服务用途给隔板进行适当改装，做成简单的服务设施，如用作有座椅和书架的阅览角。

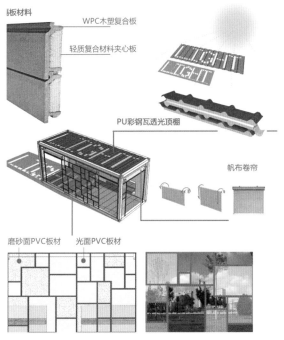

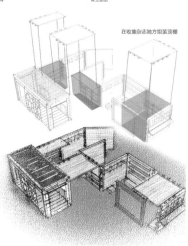

图 07 轻质材料的利用
Fig07 Using lightweight materials

图 08 遮蔽型模型
Fig08 Cover type model

查尔斯·沙（校订）
English reviewed by Charles Sands

徘徊·循迹·新生
——辽宁海城市教军山城市公园景观设计
TRACKING·WANDERING·NEWBORN
——LANDSCAPE DESIGN OF HAICHENG JIAOJUN MOUTAIN URBAN PARK IN LIAONING PROVINCE

中文标题：徘徊·循迹·新生——辽宁海城市教军山城市公园景观设计
组　　别：本科
作　　者：马斯婷
指导教师：杨立新　金 煜
学　　校：沈阳农业大学
学科专业：园林
研究方向：园林设计
分　　类：2013"园冶杯"风景园林（毕业作品、论文）国际竞赛设计类作品鼓励奖

Title: Tracking·Wandering·Newborn——Landscape Design of Haicheng Jiaojun Moutain Urban Park in Liaoning Province
Degree: Bachelor
Author: Siting Ma
Instructor: Lixin Yang, Yu Jin
University: Shenyang Agricultural University
Specialized Subject: Landscape
Topic: Landscape Design
Category: The Honorable Mention in Design Group of the 2013 "Yuan Ye Award" International Landscape Architecture Graduation Project/Thesis Competition

1. 主入口广场
2. 小空间
3. 木质廊架
4. 木平台
5. 特色种植池
6. 芳香花园
7. 木质围树椅
8. 文化主题广场
9. 小水景
10. 塑胶跑道
11. 生态养生林
12. 次入口
13. 生态防护林
14. 车行环路
15. 红色廊道
16. 专用入口
17. 健身空间
18. 儿童游乐场
19. 矿坑遗址
20. 高架木栈桥
21. 登山步道
22. 林间木栈道
23. 生态林
24. 钢架观景平台
25. 水体
26. 木栈桥
27. 膜结构
28. 休息空间

图 01 公园平面图
Fig01 Park plan

图 02 中央广场效果图
Fig02 Rendering of central square

图 03 木质景观方案
Fig03 Wooded landscape

图 04 水坑改善方案
Fig04 Pool improvements

作品简介：

本设计结合辽宁省海城市教军山原有场地的矿山和水坑而进行的。根据海城城市文化底蕴和特色，通过对矿山的探索，更新这一过程的研究，创建一个具有生态改善，娱乐休闲的多功能的城市性公园（图01）。因而此次设计就是按着徘徊·循迹·新生的主题展开而来的，通过对原有场地的利用和改造，打造了多功能的景观分区，包括金、木、水、火、土等五大项综合体现（图02-06）。本方案对矿山进行了景观化处理，利用其地势做了全园的观景平台和木质架空廊道（图07-08），并结合植物的搭配（图09），构成了景观生态相结合的城市公园（图10）。

Introduction:

The project is designed combining the mines and pools of the original site in Jiaojun Mountain, Haicheng City, Liaoning Province. Based on the profound cultural background and characteristics of the city, we explore the mine through updated research processes, and finally create an urban park with multiple functions including ecological improvement and entertainment (Fig.01).

According to the theme: "tracking, wandering, newborn", we design multi-functional landscape districts displaying an integration of metal, wood, water, fire and earth via the use and transformation of the original site (Fig.02-06). In addition, we have included some interpretive landscape elements for the mine. For example, taking advantage of the terrain, we have designed viewing platforms and wooden overhead corridors throughout the garden (Fig.07-08). Moreover, by combining the constructions with suitable plant distrubution (Fig.09), we can create an urban park that is both a functional human landscape and a thriving ecology (Fig.10).

图 05 健身娱乐区效果图
Fig05 Rendering of fitness and recreation area

图 06 矿坑景观效果图
Fig06 Rendering of pit landscape

图 07 观景平台效果图
Fig07 Rendering of viewing platform

竞赛作品 CONTEST ENTRIES

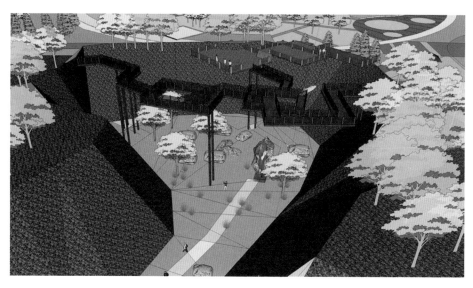

图08 架空廊道效果图
Fig08 Rendering of overhead corridor

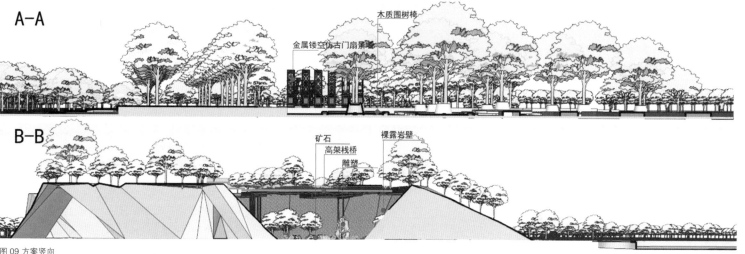

图09 方案竖向
Fig09 Vertical design

图10 公园全景图
Fig10 Panorama

查尔斯·沙（校订）
English reviewed by Charles Sands

熊猫岛
——雅安市熊猫岛主题公园景观规划与设计
PANDA ISLAND
——LANDSCAPE PLANNING AND DESIGN OF PANDA ISLAND THEME PARK IN YAAN

中文标题：熊猫岛——雅安市熊猫岛主题公园景观规划与设计	Title: Panda Island——Landscape Planning and Design of Panda Island Theme Park in Yaan
组　　别：本科	Degree: Bachelor
作　　者：周云婷	Author: Yunting Zhou
指导教师：蔡　军	Instructor: Jun Cai
学　　校：四川农业大学　现就读于西南交通大学风景园林学专业	University: Sichuan Agriculture University (Now majoring in Landscape Architecture, Southwest Jiaotong University)
学科专业：园林	Specialized Subject: Landscape Design
研究方向：景观规划与设计	Topic: Landscape Planning and Design
分　　类：2013"园冶杯"风景园林（毕业作品、论文）国际竞赛设计类作品鼓励奖	Category: The Honorable Mention in Design Group of the 2013 "Yuan Ye Award" International Landscape Architecture Graduation Project/Thesis Competition

图01 全景鸟瞰图
Fig01 Bird's eye view

竞赛作品 CONTEST ENTRIES

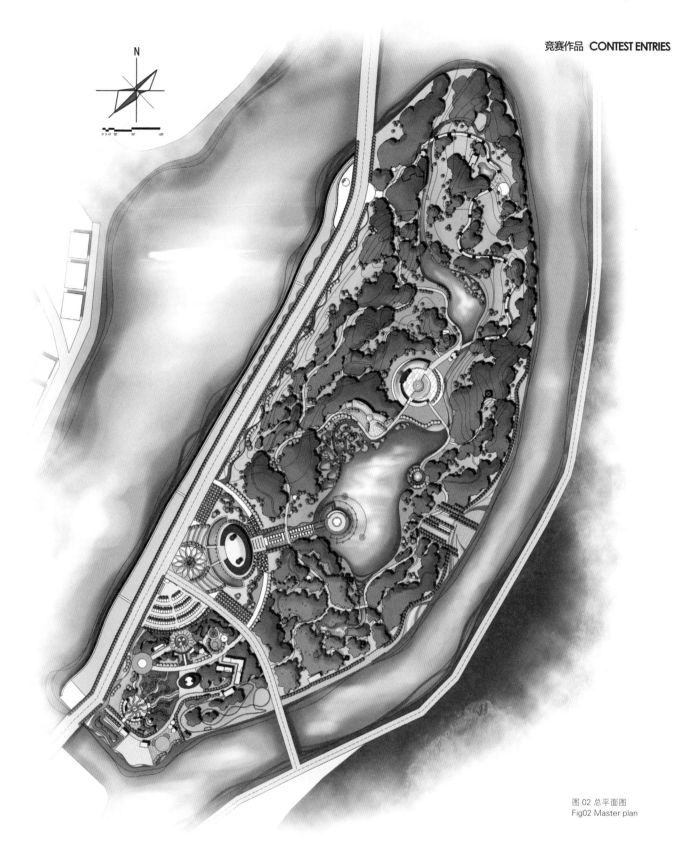

图02 总平面图
Fig02 Master plan

作品简介：

本设计主题为雅安市熊猫岛主题公园景观规划与设计，从景观层次发掘和弘扬最古老、最有价值的大熊猫文化，推动和丰富大熊猫文化、吸引国内外游客与本地居民参与大熊猫文化互动，提升雅安市民的生活质量，带动雅安社会经济更好更快地发展（图01-02）。

主入口处设置大熊猫文化馆与水上休闲空间（图03-05），模拟生境的灵感来源于海洋主题公园中的水下隧道结合地形设计，形成上中下三层空间，以人与动物不同路线为基础，让人们既可以看见动物日常生活，动物也可以自由生活在人工森林中（图06-09）。

Introduction:

By exploring the oldest and most valuable "panda culture" from the landscape level, this project aims to promote and enrich the culture of pandas to attract local residents and domestic and foreign tourists to participate in panda related activities. Consequently, it will stimulate a higher quality and faster rate of development for Yaan's economy, and thereby enhance the living quality of residents in Yaan (Fig.01-02).

At the main entrance, is the Panda culture exhibition hall and Aquatic recreational space (Fig.03-05). The basis of the simulated ecological environment is a series of underwater tunnels integrated with the terrain in the Ocean Theme Park. There are three levels of space based on the different routes of people and animals. People can catch sight of the animals' daily life while the animals are also free to live in the artificial forest (Fig.06-09).

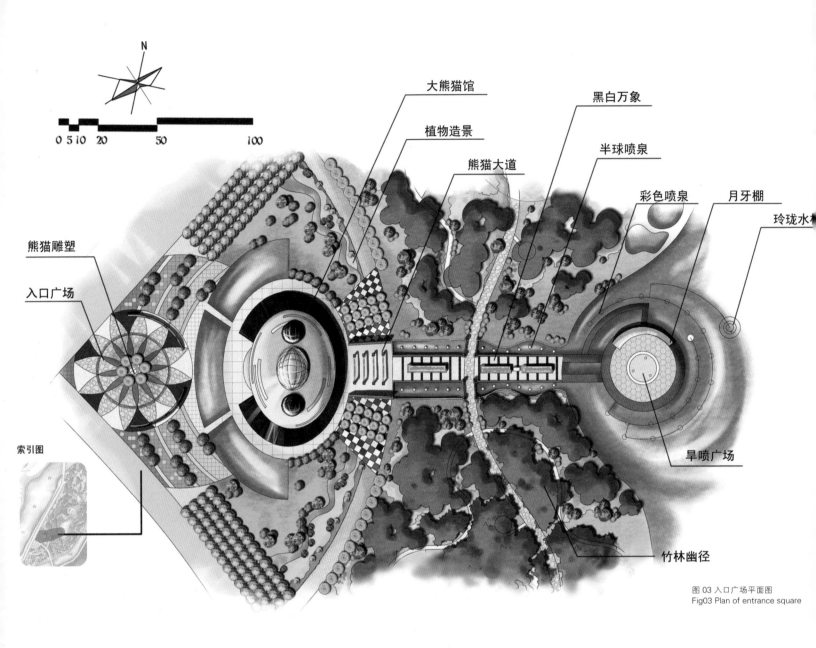

图 03 入口广场平面图
Fig03 Plan of entrance square

图 04 上善若水效果图
Fig04 Aquatic recreational space

图 05 熊猫展馆效果图
Fig05 Panda exhibition rendering

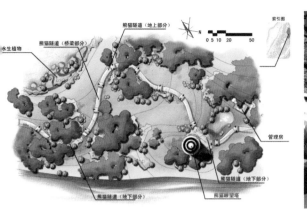

图06 模拟生境平面图
Fig06 Plan for tour of constructed ecological environment

图07 模拟生境外部效果图1
Fig07 Ecological environment tour lines rendering 1

图08 模拟生境外部效果图2
Fig08 Ecological environment tour lines rendering 2

图09 模拟生境外部效果图3
Fig09 Ecological environment tour lines rendering 3

查尔斯·沙（校订）
English reviewed by Charles Sands

图 01 总平面图
Fig01 Master plan

沁园住宅区二期景观设计
LANDSCAPE DESIGN OF THE SECOND STAGE OF QINYUAN RESIDENTIAL AREA

中文标题：沁园住宅区二期景观设计	Title: Landscape Design of the Second Stage of Qinyuan Residential Area
组　　别：本科	Degree: Bachelor
作　　者：吴竹韵	Author: Zhuyun Wu
指导教师：李春梅	Instructor: Chunmei Li
学　　校：浙江大学	University: Zhejiang University
学科专业：园林	Specialized Subject: Landscape Architecture
研究方向：居住区景观设计	Topic: Landscape Design of Residential Area
分　　类：2013"园冶杯"风景园林（毕业作品、论文）国际竞赛设计类作品鼓励奖	Category: The Honorable Mention in Design Group of the 2013 "Yuan Ye Award" International Landscape Architecture Graduation Project/Thesis Competition

图02 主入口
Fig02 Main Entrance

图03 野趣亭-1
Fig03 Pavilion with wild style-1

图04 溢香广场
Fig04 Plaza with overflowed fragrance

作品简介：

沁园位于安徽省黄山市黄山区政务新区，与10余万平方米的政务广场为邻，西邻甘芙大道，东至自然山坡，依山近水，环境优越（图01）。设计了花园洋房、水景排屋与山地别墅三类住宅，建筑与环境相互交融，住户能感受到亲水、观景、赏心的乐趣，体会"人本至上"的魅力（图02）。二期用地面积37852.1m²，总建筑面积38593.1m²，建筑占地面积9978.4m²，以多层公寓和排屋为主，绿地面积15595.0m²，绿地率达41.2%。设计注重生态，在城市中融入自然野趣（图03、08），设置休闲聚会场地（图04），设置水景，动静结合（图05-06），注重用灯光效果营造夜景，巧用宅间绿地（图07），促进邻里和谐。

Introduction:

Qinyuan is located in the new Administrative District of Huangshan, Anhui Province, next to the government plaza, It covers an area of more than 100,000 square meters. The site is bordered by natural slopes to the east and Ganfu Road to the west. It has the advantage of being surrounded by mountains and rivers (Fig.01). The project includes three types of houses: garden houses, row-houses with waterscapes, and mountain villas, with the intention of blending the structures into their corresponding environments. Residents can be close to the water and the beautiful scenery, and also experience the human-centered charm of the design (Fig.02). The second stage covers an area of 37852.1m². The total construction area is 38593.1m². The built form is comprised mainly of multi-storey apartments and row-houses with an areas of 9978.4 m². The green area covers 15595.0m² with 41.2% green coverage.

The design pays attention to ecology, which creates wildness into the urban district (Fig.03, 08). There is a plaza for relaxation and get-together for the residents (Fig.04), and create static and dynamic combination by the water scenery. Lightening at night could be used to create scenery (Fig.05-06), and the green space between houses can be used to promote community harmony(Fig.07).

图 05 组合水景
Fig05 Combined waterscape

图 06 组合水景夜景
Fig06 Combined waterscape at night

图 07 邻里天地（宅间绿地）
Fig07 Neighborhood (green space between houses)

图 08 野趣亭 -2
Fig08 Pavilion with wild style-2

查尔斯·沙（校订）
English reviewed by Charles Sands

山水比德
SUN&PARTNERS INCORPORATION

PROFESSION CHANGE LIFE

广州山水比德景观设计有限公司

湘江一号实景图

广州珠江新城·珠江北岸 文化创意码头
领航创意 共创未来

广州市天河区珠江新城临江大道685号红专厂F19栋
TELL：020-37039822　37039823　37039825
FAX：020-37039770
E-MAIL：SSBD-S@163.COM

全国招募

TEL：020-37039822
FAX：020-37039770

服务对象：

《世界园林》征稿启事
Notes to Worldscape Contributors

1. 本刊是面向国际发行的主题性双语（中英文）期刊。设有作品实录、专题文章、人物／公司专栏、热点评论、构造、工法与材料（含植物）5个主要专栏。与主题相关的国内外优秀作品和文章均可投稿。稿件中所有文字均为中英文对照。所有投稿稿件文字均为Word文件。作品类投稿文字中英文均以1000-2500字为宜，专题文章投稿文字的中英文均以2500-4000字为宜。

2. 来稿书写结构顺序为：文题（20字以内，含英文标题）、作者姓名（中国作者含汉语拼音，外国作者含中文翻译）、文章主体、作者简介（包括姓名、性别、籍贯、最高学历、职称或职务、从事学科或研究方向、现供职单位、所在城市、邮编、电子信箱、联系电话）。作者两人以上的，请注明顺序。

3. 文中涉及的人名、地名、学名、公式、符号应核实无误；外文字母的文种、正斜体、大小写、上下标等应清楚注明；计量单位、符号、号字用法、专业名词术语一律采用相应的国家标准。植物应配上准确的拉丁学名。扫描或计算机绘制的图要求清晰、色彩饱和，尺寸不小于15cm×20cm；线条图一般以A4幅面为宜，图片电子文件分辨率不应小于300dpi（可提供多幅备选）。数码相机、数码单反相机拍摄的照片，要求不少于1000万像素（分辨率3872×2592），优先使用jpg格式。附表采用"三线表"，必要时可适当添加辅助线，表格上方写明表序和中英文表名，表序应于内文相应处标明。

4. 作品类稿件应包含项目信息：项目位置／项目面积／委托单位／设计单位／设计师（限景观设计）／完成时间。

5. 介绍作品的图片（有关设计构思、设计过程及建造情况和实景等均可）及专题文章插图均为jpg格式。图片请勿直接插在文字文件中，文字稿里插入配图编号，文末列入图题（须含中英对照的图号及简要说明）。图片文件请单独提供，编号与文字文件中图号一致。图题格式为：图01 xxx/Fig 01xxx。图片数量15-20张为宜。可标明排版时对图片大小的建议。

6. 文稿一经录用，即每篇赠送期刊2本，抽印本10本。作者为2人以上，每人每篇赠送期刊1本，抽印本5本。

7. 投稿邮箱：worldscape_c@chla.com.cn 联系电话：86-10-88364851

---

1. Worldscape is an international thematic bilingual journal printed in dual Chinese and English. It covers five main columns including Projects, Articles, Masters / Ateliers, Comments, and Construction & Materials (including plants). The editors encourage the authors to contribute projects or articles related to the theme of each issue in both Chinese and English. All submissions should be submitted in Microsoft Word (.doc) format. Chinese articles should be 1000-2500 characters long. English articles should be 2500-4000 words long.

2. All the submitted articles should be organized in the following sequence: title (no more than 20 characters and the English title should be contained); author's name (for Chinese authors, pin yin of the name should be accompanied; for foreign authors, the Chinese translation of the name should be accompanied if applicable); main body; introduction to the author (including name, gender, native place, official academic credentials, position/title, discipline/research orientation, current employer, city of residence, postal code, E-mail, telephone number). For articles written by two or more authors, please list the names in sequence.

3. All persons, places, scientific names, formulas and symbols should be verified. The English submissions should be word-processed and carefully checked. Measuring units, symbols, and terminology should be used in accordance with corresponding national standards. Plants should be accompanied with correct Latin names. Scanned or computer-generated pictures should be sharp and saturated, and the size should be not less than 15cmx20cm. Diagrams and charts should be A4-sized. The resolution of digital images should be not less than 300dpi (authors are encouraged to provide a selection of images for the editors to choose from). The resolution of pictures generated by digital camera and digital SLR camera should be not less than 3872x2592, and .jpg formatted pictures are preferred. Annexed tables should be three-lined, and if necessary, auxiliary lines may be used. All tables should be sequenced and correspond to the text. Chinese-English captions should be contained.

4. All the submitted materials should be accompanied with short project information: site, area, client, design studio (atelier or company name), landscape designers (landscape architects) and completion date (year).

5. All project images (to illustrate the concept, design process, construction and built form) should be .jpg format. The images should be sent separately and not integrated in the text. All images should be numbered, and the numbers should be represented in the main body of the text. At the end of the text, captions and introductions to the images should be attached (Chinese-English bilingual text). The caption should be formatted as Fig 01 xxx. No more than 20 images should be submitted. Suggestions to image typeset may be attached.

6. The author of each accepted article will be sent 2 copies of the journal and 10 copies of the offprint. In the case of articles with 2 or more authors, each author will be sent 1 copy of the journal and 5 offprints.

7. Articles should be submitted to: worldscape_c@chla.com.cn Tel: 86-10-88364851

青草地园林市政

——营建城市新型绿地　探寻花卉发展模式

浙江青草地园林市政建设发展有限公司是具有国家壹级城市园林绿化企业资质、市政公用工程施工总承包贰级、绿化造林施工资质乙级、绿化造林设计资质乙级、园林古建筑工程专业承包叁级、河湖整治工程专业承包叁级、城市及道路照明工程专业承包叁级、体育场地设施工程专业承包叁级、机电设备安装工程专业承包叁级，集园林绿化设计施工、园林植物科学研究、花卉生产销售、园林信息咨询和鲜花礼仪服务一体的综合型园林市政企业。

湖州潜山公园景

海天公

海天公园假山景观

湖州潜山公园景

中华树艺苑

  公司确定以"质量立业"为发展定位，辨证地处理量的跨越和质的提高的关系。先后承接了海天公园（包括海天高尔夫球场）整体绿化工程、杭州樱花小筑室外景观工程、萧山区风情大道北伸绿化工程、丽水市滨江景观带工程、红谷滩新区总体绿化工程、南昌市象湖公园景观工程、湖州潜山公园景观工程、湖州南浔嘉沁园室外总布景观绿化工程、湖州大剧院景观工程、江苏省太仓港口开发区管理委员会、绿城•桂花园一起景观绿化工程、上海大上海会德丰广场硬景工期、上海新华路一号景观工程、银亿海尚广场、乐清东山公园一期建设工程、上海三甲港江畔御庭别墅绿化工程、温州广汇景园绿化工程等400余项，多项工程获得过"杜鹃花奖"、"百花奖"、"茶花杯"、"最佳人居住环境奖"、"飞英杯"、浙江省优秀园林"金奖"等奖项，绿化市场逐渐向全国拓展。

和景中象城一期景观

温州广汇景观

独特的视角
创新的设计

城市规划·景观设计·建筑设计·项目策划

地址：北京市朝阳区北苑路甲13号北辰新纪元大厦2座1705室
电话：010-8492 1962/ 8492 7362   传真：010-8482 8780
邮箱：marketing@macromind.cn   网站：www.macromind.cn